Synthesis Lectures on Mathematics & Statistics

Series Editor

Steven G. Krantz, Department of Mathematics, Washington University, Saint Louis, MO, USA

This series includes titles in applied mathematics and statistics for cross-disciplinary STEM professionals, educators, researchers, and students. The series focuses on new and traditional techniques to develop mathematical knowledge and skills, an understanding of core mathematical reasoning, and the ability to utilize data in specific applications.

Yaiza Canzani · Jeffrey Galkowski

Geodesic Beams
in Eigenfunction Analysis

 Springer

Yaiza Canzani
Department of Mathematics
University of North Carolina at Chapel Hill
Chapel Hill, NC, USA

Jeffrey Galkowski
Department of Mathematics
University College London
London, UK

ISSN 1938-1743 ISSN 1938-1751 (electronic)
Synthesis Lectures on Mathematics & Statistics
ISBN 978-3-031-31588-6 ISBN 978-3-031-31586-2 (eBook)
https://doi.org/10.1007/978-3-031-31586-2

This Springer imprint is published by the registered company Springer Nature Switzerland AG
The registered company address is: Gewerbestrasse 11, 6330 Cham, Switzerland

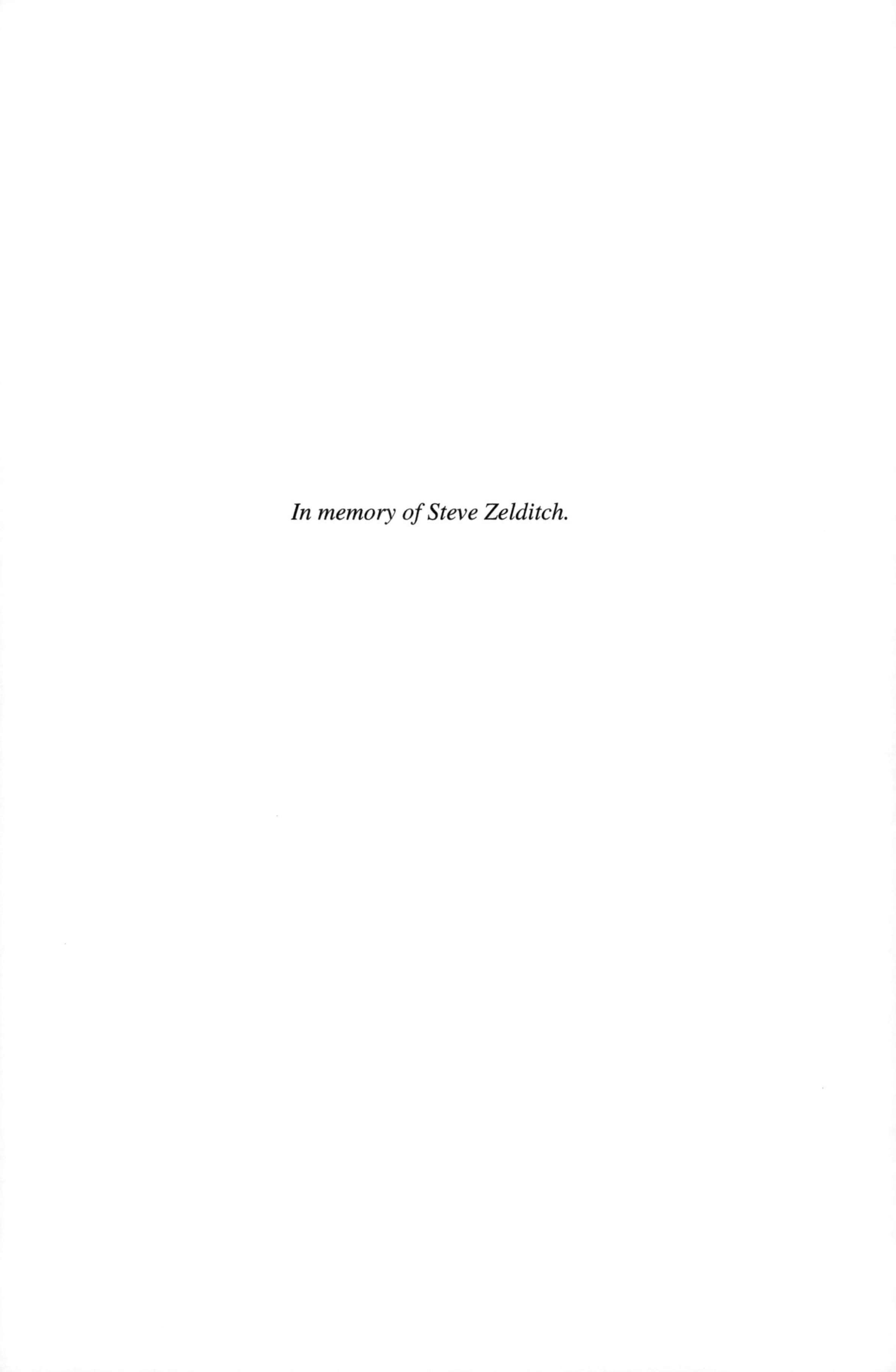

In memory of Steve Zelditch.

Preface

This book aims to explain the concepts behind the geodesic beam method that we have developed to study the behavior of high energy eigenfunctions. The idea for geodesic beams was inspired by the work of Koch-Tataru-Zworski **[KTZ07]** as well as the second author's work with Toth **[GT17]**. Geodesic beams originally appeared in **[Gal19]** and were developed into their current form in **[CG19b, CG21, CG19a, CG20b]**.

The book is intended to be accessible to graduate students and is aimed at readers interested in studying the behavior of eigenfunctions. Although familiarity with semiclassical analysis will be helpful when reading this text, we provide a quick, axiomatic introduction to the subject in Chap. 3. This chapter, in principle, contains all of the semiclassical analysis results necessary to understand the rest of the text. For more detailed treatments of the semiclassical machinery, we refer the reader to the books **[Zwo12, DS99, DZ19]**.

Throughout the book, we chose to present full proofs when they illustrate important ideas and are central to our method. We instead present outlines of proofs when the technical details can obscure the ideas. In this case, we refer the reader to the corresponding article for the full details of the proof. We hope that our choices will make the text relatively easy to read, while still containing the main points of our analysis.

Our presentation of various parts of this text was heavily influenced by Zworski's book **[Zwo12]** from which the authors first learned semiclassical analysis. Our approach to semiclassical analysis was informed by long interactions with Maciej Zworski, Semyon Dyatlov, and Andras Vasy, and our perspective on the analysis of eigenfunctions has been guided by Dima Jakobson, Peter Sarnak, John Toth, and Steve Zelditch. This text would not have been possible without their support.

Y. Canzani was supported by the Alfred P. Sloan Foundation, NSF CAREER Grant DMS-2045494, and NSF Grant DMS-1900519. J. Galkowski acknowledges support from the EPSRC through grants EP/V001760/1 and EP/V051636/1.

Chapel Hill, USA
London, UK

Yaiza Canzani
Jeffrey Galkowski

Contents

Introduction 1

The goal of this text is to explain the ideas behind the authors' work on the behavior of Laplace eigenfunctions on a compact Riemannian manifold via decompositions into what we call geodesic beams.

The heart of the geodesic beam method involves a decomposition of a function, u, into building blocks supported in semiclassically small sets. Indeed, in general, when trying to understand the behavior of u, it is natural to decompose it into building blocks that are more easily understood. To this end, it is often helpful to use the 'smallest' building blocks into which u can be decomposed. Here, because we intend to work with functions oscillating at approximately frequency λ, 'smallest' is dictated by the uncertainty principle.

We start by reviewing two standard unstructured types of localizers: coherent state localizers and pointwise localizers. These, however, turn out not to be well adapted for the study of Laplace eigenfunctions. Afterward, we present the concept of structured localizers which almost preserve (at least locally) the property of being an eigenfunction. In particular, we present the basis for a decomposition of a function into what we call geodesic beams.

Unstructured localizers are appealing because they do not depend on the fine structure of the problem under consideration. We start by discussing two such families of localizers on $\mathbb{R}^n$.

Coherent state decomposition: Using the Fourier–Bros–Iagolnitzer transform (or a wavelet type decomposition) one can decompose any function as

$$u = \sum_j a_j u_j, \qquad u_j(x) = \lambda^{\frac{n}{4}} e^{-\lambda|x-x_j|^2/2} e^{i\lambda\langle x-x_j, \xi_j\rangle}, \tag{1.1}$$

where $\lambda > 0$ and (x_j, ξ_j) are points in the standard $\sqrt{\frac{\pi}{\lambda}}$ lattice for $T^*\mathbb{R}^n \approx \mathbb{R}^{2n}$ [Chr16].

© The Author(s), under exclusive license to Springer Nature Switzerland AG 2023
Y. Canzani and J. Galkowski, *Geodesic Beams in Eigenfunction Analysis*, Synthesis Lectures on Mathematics & Statistics, https://doi.org/10.1007/978-3-031-31586-2_1

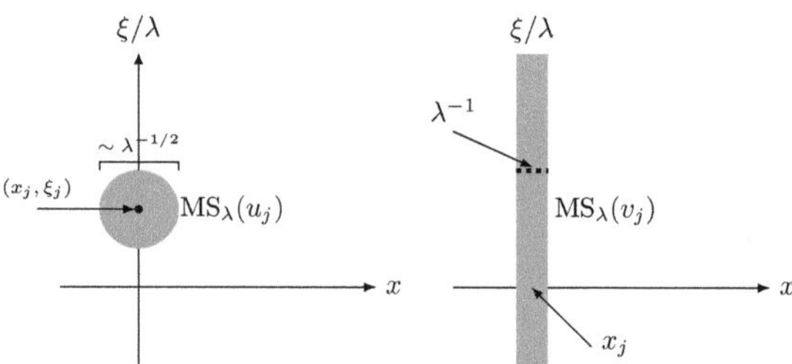

Fig. 1.1 This figure illustrates the two different behaviors of the building blocks in each of the unstructured decomposition described here. Shown in the picture are the semiclassical wavefront sets of the unstructured localizers. See Definition 3.2.1 and Remark 3.2.8

Since we are interested in functions, u, that oscillate at frequency $\sim \lambda$, it is natural to rescale the Fourier variable by λ^{-1} so that the Fourier transform, in the new variables, lies in a compact set. We then observe that the coherent state u_j is nearly supported in the ball $B(x_j, C\lambda^{-1/2})$ and have (rescaled) Fourier support in the ball $B(\xi_j, C\lambda^{-1/2})$. Thus, each u_j is supported in position–frequency space on a set with volume λ^{-n}, the smallest such volume allowed by the uncertainty principle.

Note that this decomposition treats all position and momentum directions uniformly. See Fig. 1.1 for a picture of the wavefront set of one such function. Notice that there is $c > 0$ such that for all j,

$$\|(-\lambda^{-2}\Delta - 1)u_j\|_{L^2} \geq c\lambda^{-\frac{1}{2}}\|u_j\|_{L^2},$$

and, because of this, even if u solves $(-\lambda^{-2}\Delta - 1)u = 0$, each localizer u_j does not behave like a local solution for the eigenvalue problem.

Our next set of localizers treats position and momentum differently.

Pointwise localizers: If, instead of treating all directions in the same way, one views the physical space as the most important localization, we are led to decompose

$$u \sim \sum_j v_j, \qquad v_j(x) = \chi(\lambda(x - x_j))u(x), \tag{1.2}$$

for x_j in the standard λ^{-1} lattice on $\mathbb{R}^n$ and with $\chi \in C_c^\infty(\mathbb{R}^n)$. Then, v_j is the 'smallest' piece one can decompose a function into in the sense that it is maximally localized in position at the uncertainty principle scale; that is it is supported in $B(x_j, C\lambda^{-1})$ and its (rescaled) Fourier transform is supported in $B(0, C)$. In particular, each one is supported in position–frequency space on a set with volume λ^{-n}, the smallest such volume allowed by the uncertainty principle. From the point of view of eigenfunctions, these pointwise localizers behave even worse than the coherent states. Indeed, for an eigenfunction u we have

$$\|(-\lambda^{-2}\Delta - 1)v_j\|_{L^2} \geq c\|u\|_{L^2(B(0,C\lambda^{-1}))}.$$

In particular, each localizer v_j does not behave like a local solution for the eigenvalue problem.

Unstructured localizers do not allow us to take advantage of the structure which comes from the fact that u solves the equation

$$Pu = 0, \qquad P := -\lambda^{-2}\Delta - I.$$

Indeed, neither the coherent state localizer nor the pointwise localizers has the property that any individual member, u_j of the decomposition $u \sim \sum u_j$ satisfies the equation $Pu_j \sim 0$ even in some small open set. Since, in many cases, it is precisely this structure that we wish to exploit, it is natural to work with building blocks that respect the equation. Another feature that will be useful for our localizers is that they arise as pseudodifferential partitions of unity (albeit in potentially exotic classes of pseudodifferential operators). (see Chap. 3 for a quick review of semiclassical pseudodifferential operators.)

We now turn our attention to the Laplace equation on a compact, Riemannian manifold (M, g), writing

$$P := -\lambda^{-2}\Delta_g - I, \qquad \text{with symbol} \qquad p(x, \xi) := |\xi|_g^2 - 1.$$

In order to understand how our localizers could respect the equation that u satisfies, given a pseudodifferential cutoff X, we consider the features of X that allow it to respect the structure of a solution to the equation $Pu = 0$. To do this, we want at least locally to have $PXu \sim 0$. This is only possible if

$$0 \sim PXu = XPu + [P, X]u = [P, X]u,$$

so X must at least approximately commute with P. Since X is pseudodifferential, this means that, at the level of principal symbols, $\sigma([P, X]) \sim 0$. In other words, $H_p\sigma(X) \sim 0$, and so the symbol of X must be, at least locally, invariant under the Hamiltonian flow for p. Here, H_p denotes the Hamiltonian vector field induced by the symbol p. Since $p = |\xi|_g^2 - 1$, the Hamiltonian flow is the geodesic flow induced by the metric g.

We now discuss our main tool: geodesic beam localizers (see Chap. 6 for a detailed description). These are, in a sense, the smallest possible structured localizers that respect the equation $Pu = 0$.

Geodesic beam localizers: The analog of the coherent state decomposition, when one wants to respect the structure imposed by P, are geodesic beam decompositions; which take the form

$$u \sim \sum_j G_j u, \tag{1.3}$$

where G_j is a pseudodifferential operator with symbol supported in a $\sim \lambda^{-1/2}$ neighborhood of an approximately unit length piece of geodesic running through a point $(x_j, \xi_j) \in S^*M$. [1]

We call the functions $G_j u$ geodesic beams. They are localized to a set with volume $\sim \lambda^{-n}$ in position–momentum space (and hence at the maximum scale allowed by the uncertainty principle) and, at least locally, preserve $Pu = 0$. Indeed, for $\varepsilon > 0$ small enough, (See Fig. 1.2 for a picture of the support of these cutoffs.)

$$\|PG_j u\|_{L^2(B(x_j,\varepsilon))} = O(\lambda^{-\infty})\|u\|_{L^2},$$

and hence G_j remains a strong quasimode for P near the point x_j. Because of this fact, it will be relatively easy to prove optimal estimates on $G_j u$ and hence to study u via this geodesic beam decomposition. The purpose of this text is to construct the geodesic beam decomposition and study the behavior of eigenfunctions by understanding the behavior of geodesic beams.

When we turn to the study of L^p norms for eigenfunctions in Sect. 2.2.4 we will also use a structured analog of the pointwise localizers.

Zonal harmonic type localizer: The analog of the pointwise localizer above when one wants to respect the structure imposed by P, are what we call zonal harmonic type localizers. They take the form

$$Zu, \tag{1.4}$$

where Z is supported in an $\sim \lambda^{-1}$ neighborhood of the unit flow-out of T_x^*M by $\exp(t H_p)$ for some $x \in M$. [2] The localizers Z are very far from the pointwise localizers, $\chi(\lambda x)$. In particular, they have support on $O(1)$ sized physical sets. However, they are maximally localized along flow-lines for H_p which pass near x. Because of this, they are very effective for controlling from both above and below the behavior of u at a single point. The apparent down-side of this decomposition is that the cutoff's Z must be carefully constructed in a second-microlocal calculus. However, we will see when discussing L^p norms, that this 'down'-side is actually an essential feature when studying eigenfunctions.

The main reason why geodesic beam localizers prove to be exceptionally useful among structured localizers is that they are the only 'smallest' structured localizer for which an obvious decomposition of a function exists. Indeed, in order to localize to a higher dimensional (locally) H_p invariant set, Γ, at the uncertainty principle scale, one must use pseudodifferential calculi adapted to Γ. Since, for $\Gamma_1 \neq \Gamma_2$, the calculus associated to Γ_1 and that associated to Γ_2 are typically incompatible it is difficult to decompose a function as a sum over this type of localizer. For the zonal harmonic type localizer associated to a point x_i, we use the set

---

[1] Below, we will actually take an $\lambda^{-\delta}$ neighborhood of the geodesic with $0 < \delta < \frac{1}{2}$, since it will not be necessary to localize maximally.

[2] Below, we will actually take an $\lambda^{-\rho}$ neighborhood of the flow-out with $\frac{1}{2} < \rho < 1$, since it will not be necessary to localize maximally.

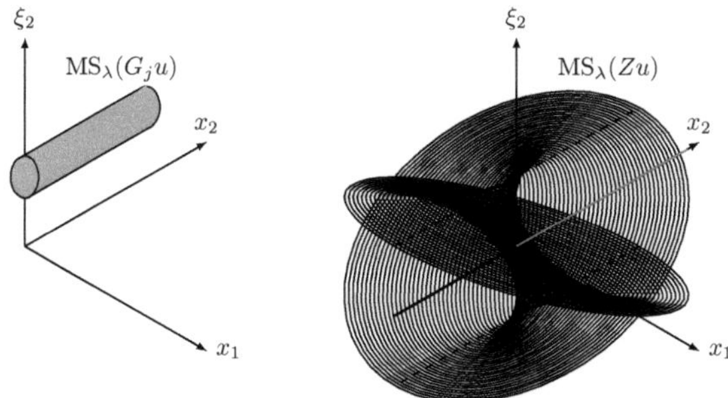

Fig. 1.2 Semiclassical wavefront sets of structured localizers projected into the (x, ξ_2) variables. The geodesic beam $G_j u$ is localized to a tube built around a geodesic running in the x_2 direction. See Definition 3.2.1 and Remark 3.2.8. The Zonal harmonic type localizer is localized to Γ_i with $x_i = 0$ for Γ_i as in (1.5)

$$\Gamma_i = \left\{ \exp(t H_{|\xi|_g})(x_i, \xi) \; : \; \xi \in T^*_{x_i} M, \; \tfrac{1}{2} \le |\xi|_{g(x_i)} \le 2, \; |t| \le 1 \right\}. \tag{1.5}$$

(See Fig. 1.2 for a picture of Γ_i.)

Throughout this text, we focus mainly on geodesic beams as result; with zonal localizers playing a role in our analysis of L^p norms. We hope that this text will encourage the use of structured localizers.

1.0.1. Structure of the Book.

The book starts in Chap. 2 with a brief introduction to the Laplace operator including motivation for the study of eigenfunctions. Furthermore, we recall the state of the art estimates on eigenfunctions and eigenvalues. In particular, we state a variety of theorems proved using geodesic beams: estimates on the eigenfunctions' L^∞ norms, averages over submanifolds, and L^p norms, and on the remainder for the Weyl law.

In Chap. 3, we provide an axiomatic introduction to semiclassical analysis. We introduce semiclassical pseudodifferential operators and the corresponding calculus on manifolds. This chapter contains the main concepts and theorems from semiclassical analysis, largely without proof. We recommend that the reader unfamiliar with semiclassical analysis treat this chapter as a black box which gives the rules of the semiclassical calculus. We refer the interested reader to [Zwo12, DZ19] for detailed introductions to the subject.

Chapter 4 gives the basic properties of Laplace eigenfunctions. Chapter 5 recalls the method used by Koch–Tataru–Zworski [KTZ07] to prove the Hörmander L^∞ estimate on Laplace eigenfunctions. The results in this section will be modified and adapted to obtain sharp estimates on the L^∞ norms of geodesic beams.

Chapter 6 contains the construction of geodesic beam decompositions as well as the main technical estimates used in understanding these decompositions. We also introduce the concepts of good covers and partitions which appear throughout our analysis of eigenfunctions.

Chapter 7 is dedicated to the application of the geodesic beam framework to control on the Laplace eigenfunctions' L^∞ norms, averages over submanifolds, and L^p norms, as well as on the remainder term for the Weyl law. This control is expressed in terms of the behavior of families of geodesic adapted to each problem.

Finally, Chap. 8 sketches the main dynamical ideas that are used to prove the estimates in concrete geometric settings in Chap. 2 using the theorems in Chap. 7. That is, under various dynamical assumptions we explain how to build appropriate families of geodesic beams which yield quantitative improvements on the standard bounds for eigenfunction concentration and Weyl Laws.

The Laplace Operator

2

We work on a smooth, compact, connected Riemannian manifold (M, g) of dimension n. Throughout most of the text, we assume that the manifold has no boundary.

The Laplace-Beltrami operator, or Laplacian, on (M, g) is defined as

$$\Delta_g = \mathrm{div}_g \nabla_g \tag{2.1}$$

where div_g is the metric divergence operator and ∇_g is the metric gradient. A priori, this operator is defined with domain equal to $C^\infty(M)$.

In any local coordinates $(x_1, \ldots, x_n)$ on M the Laplacian takes the form

$$\Delta_g = \frac{1}{\sqrt{|\det g|}} \sum_{i,j=1}^{n} \partial_{x_i}\left(g^{ij}\sqrt{|\det g|}\partial_{x_j}\right). \tag{2.2}$$

Here, $\det g$ denotes the determinant of the matrix representation $(g_{ij})_{i,j=1}^{n}$ for g in the given coordinates. It is easy to check that when the metric is flat, like in $\mathbb{R}^n$, the Laplacian simplifies to

$$\Delta = \sum_{i=1}^{n} \partial_{x_i}^2.$$

In analogy to what happens in $\mathbb{R}^n$, one may think of the Laplacian on (M, g) as the operator that acts by differentiating twice with respect to each direction. Indeed, let us fix a point $x \in M$ and let $\gamma_1, \ldots, \gamma_n : [-1, 1] \to M$ be n geodesics in (M, g) with $\gamma_i(0) = x$ for all i. Suppose further that the geodesics are orthogonal at x. That is, $\langle \dot\gamma_i(0), \dot\gamma_j(0)\rangle_g = 0$, when $i \neq j$. Then,

$$\Delta_g f(x) = \sum_{i=1}^{n}(f \circ \gamma_i)''(0).$$

Y. Canzani and J. Galkowski, *Geodesic Beams in Eigenfunction Analysis*, Synthesis Lectures on Mathematics & Statistics, https://doi.org/10.1007/978-3-031-31586-2_2

The Laplacian is a symmetric and positive definite operator. We will see in Chap. 4 that it can be extended to a self-adjoint operator acting on $L^2(M)$ whose domain is $H^2(M)$. We will also prove the Spectral Theorem for the Laplacian. In order to state it, we record that, from now on, we write $L^2(M)$ to mean $L^2(M, \mathrm{dv}_g)$ where dv_g is the volume form induced by the Riemannian metric. In particular, in local coordinates $(x_1, \ldots, x_n)$, we have $\mathrm{dv}_g = \sqrt{\det g}\, dx_1 \wedge \cdots \wedge dx_n$.

Theorem 2.0.1 (Spectral Theorem) *There exists a complete orthonormal basis $\{u_{\lambda_j}\}_j$ of $L^2(M)$ consisting of C^∞ eigenfunctions of Δ_g with*

$$\Delta_g u_{\lambda_j} = -\lambda_j^2 u_{\lambda_j}, \qquad 0 = \lambda_1 \leq \lambda_2 \leq \ldots \ \to \infty.$$

2.1 Why Study Laplace Eigenfunctions and Eigenvalues?

In this section we describe various natural phenomena in which the Laplace operator plays a crucial role. We briefly motivate the relevance of studying Laplace eigenvalues and eigenfunctions in quantum mechanics, heat propagation, wave propagation and inverse problems.

Although in the rest of the text we will assume that (M, g) has no boundary, we will drop this assumption in this section with the purpose of motivating the study of eigenfunctions and eigenvalues. If M has a boundary, ∂M, we will assume it is smooth. We will say that Dirichlet boundary conditions are imposed on a Laplace eigenfunction, φ, whenever we ask that $\varphi(x) = 0$ for $x \in \partial M$.

2.1.1 Quantum Mechanics

If one wishes to understand the behavior of a free quantum particle on (M, g) (under the assumption that there are no external forces), then one needs to solve the *Schrödinger Equation*

$$-\frac{\hbar^2}{2m} \Delta u(x, t) = i\hbar \partial_t u(x, t),$$

where $\hbar$ is Planck's constant and m is the mass of the free particle. Normalizing u so that $\|u(\cdot, t)\|_{L^2} = 1$ one interprets $u(x, t)$ as a probability density. That is, if $A \subset \Omega$ then the probability that the free quantum particle be inside A at time t is given by $\int_A |u(x, t)|^2 \mathrm{dv}_g(x)$.

The general solution to the Schrödinger Equation can be expressed via an orthornormal basis of $L^2(M)$ made of Laplace eigenfunctions, $-\Delta u_{\lambda_k} = \lambda_k^2 u_{\lambda_k}$, (see Theorem 2.0.1) as

$$u(x, t) = \sum_k a_k e^{i\lambda_k^2 t} u_{\lambda_k}(x)$$

where the coefficients a_k are chosen depending on the initial conditions. In particular, the probability for finding a free quantum particle of energy λ_k^2 inside the set $A \subset M$ is given by the integral $\int_A |u_{\lambda_k}(x)|^2 dv_g(x)$.

2.1.2   Heat Propagation

Heat propagation in (M, g) is dictated by the *Heat Equation*

$$\Delta u(x, t) = \partial_t u(x, t), \tag{2.3}$$

Here, the solution $u(x, t)$ represents the temperature at the point $x \in M$ at time t.

The general solution to the heat equation can again be expressed via an orthornormal basis of $L^2(M)$ made of Laplace eigenfunctions, $-\Delta u_{\lambda_k} = \lambda_k^2 u_{\lambda_k}$, (see Theorem 2.0.1) as

$$u(x, t) = \sum_k a_k e^{-\lambda_k^2 t} u_{\lambda_k}(x), \qquad x \in M, \tag{2.4}$$

where the coefficients a_k are chosen depending on the initial conditions.

If M is an insulated region and one observes an initial heat distribution on its edge, ∂M, then one could ask what the temperature distribution inside M will be after a long enough period of time (that is, the steady state temperature distribution). To answer this question, one needs to find a solution of the heat equation that is independent of time. The steady state temperature solution will be a function $u(x, t)$ such that $\Delta u = 0$. In this case, u is said to be a harmonic function in x.

2.1.3   Wave Propagation

Waves propagate accross (M, g) according to the *Wave equation*

$$\Delta u(x, t) = \partial_t^2 u(x, t), \tag{2.5}$$

where $u(x, t)$ denotes the height of the wave above the point x at time t.

One could think of M as the membrane of a drum, in which case its boundary $\partial \Omega$ would be attached to the rim of the drum. If one wishes to study the vibration generated once the drum is hit, the one should also solve the wave equation, but this time with boundary condition $u(x, t) = 0$ for all $x \in \partial M$.

The general solution to the wave equation can be expressed via an orthonormal basis of $L^2(M)$ made of Laplace eigenfunctions, $-\Delta u_{\lambda_k} = \lambda_k^2 u_{\lambda_k}$, (see Theorem 2.0.1) as

$$u(x, t) = \sum_k a_k e^{i\lambda_k t} u_{\lambda_k}(x), \tag{2.6}$$

where the coefficients a_k are chosen depending on the initial conditions. When working on the drum membrane scenario, the boundary condition would impose that $u_{\lambda_k}(x) = 0$ for $x \in \partial M$.

One application of interest is whether, by applying a forcing term in some open set, $\Omega \subset M$, the solution to the wave equation can be driven to a given state u_0 or whether the solution to the heat equation can be driven to 0. Because of the intimate relationship between Laplace eigenfunctions and the solutions of the wave equation (see (2.6)), these problems are nearly equivalent to certain lower bounds on the L^2 norm, $\|u_{\lambda_j}\|_{L^2(\Omega)}$. Indeed, contollability for the wave equation is deeply connected to the question whether $\inf_j \|u_{\lambda_j}\|_{L^2(\Omega)} > 0$ while the controllability of the wave equation is closely related to the lower bound $\|u_{\lambda_j}\|_{L^2(\Omega)} > ce^{-C\lambda_j}$ for all j.

2.1.4 Inverse Problems

Laplace eigenvalues encode a great deal of information on the geometry of (M, g). For example, if $M \subset \mathbb{R}^2$ and Dirichlet boundary conditions are imposed on ∂M, then the eigenvalues of the Laplacian, $\lambda_1 \leq \lambda_2 \leq \ldots$, satisfy that, as $t \to 0^+$,

$$\sum_{k=1}^{\infty} e^{-\lambda_k t} \sim \frac{1}{4\pi t} \left(\text{area}(M) - \sqrt{4\pi t} \, \text{length}(\partial M) + \frac{2\pi t}{3}(1 - \gamma(M)) \right),$$

where $\gamma(M)$ is the number of holes of M. The first term was proved by Weyl in 1911. The second term was proved in 1954 by Pleijel. It shows that you can hear whether a drum is circular or not (because of the isoperimetric inequality, $[\text{length}(\partial M)]^2 \geq 4\pi \text{area}(M)$, whose equality is only attained by circles).

It is only natural to then ask the following question. If you have perfect pitch, could you derive the shape of a drum from the music you hear from it? More generally, can you determine the structural soundness of an object by listening to its vibrations? This question was first posed by Schuster in 1882 and is now a famous question, often referred to as *Can you hear the shape of a drum?*. Unfortunately, the answer to the question is no. This was proved in 1992 by Gordon, Webb and Wolpert [GWW92]; because of their work, we now know many planar domains which are not isometric, but have exactly the same spectrum.

2.2   Statement of the State of the Art Estimates

Eigenfunction concentration and eigenvalue distribution are topics that have been extensively studied over the last century. In this section we introduce the known bounds for estimates on sup-norms, averages over submanifolds, error terms in the Weyl Law for the eigenvalue counting function, and estimates for L^p-norms of eigenfunctions. This is done in Sects. 2.2.1, 2.2.2, 7.4, and 7.3 respectively.

2.2.1 L^∞ Norms

Let u_λ be a Laplace eigenfunction of eigenvalue λ^2. Beginning in the 1950's, the works of Levitan, Avakumović, and Hörmander [Lev52, Ava56, Hör68] proved the estimate

$$\|u_\lambda\|_{L^\infty} = O(\lambda^{\frac{n-1}{2}}), \tag{2.7}$$

as $\lambda \to \infty$; known to be saturated on the round sphere. This bound was improved to $o(\lambda^{\frac{n-1}{2}})$ by Safarov,Sogge, Toth, Zelditch and the second author [Saf88, SZ02a, STZ11, SZ16a, SZ16b, GT17, Gal19] under various dynamical assumptions at x. Notably, [Saf88, SZ02a] were the first to study L^∞ bounds under purely local dynamical assumptions. In addition to these improvements to $o(\lambda^{\frac{n-1}{2}})$, when (M, g) has no conjugate points, a quantitative improvement of the form

$$\|u_\lambda\|_{L^\infty} = O(\lambda^{\frac{n-1}{2}}/\sqrt{\log \lambda}),$$

as $\lambda \to \infty$, has been known since the classical work of Bérard [Bér77a, Bon17a, Ran78]. However, until the authors' work [CG21, CG19a] developing the tool of geodesic beams, no quantitative improvements were available without *global geometric* assumptions on (M, g). (See the introduction Chap. 1 for a rough introduction to the idea of geodesic beams and Chap. 6 for their construction.)

The L^∞ bound in (2.7) is saturated on the round sphere, S^n, if one chooses u_λ to be a zonal harmonic that peaks at the given point $x \in S^n$ (see Fig. 2.1). This phenomenon is intimately related to the fact that all geodesics through x are closed. In addition, on the sphere every point is maximally self-conjugate. In general, a point $x \in M$ is said to be conjugate to $y \in M$ if there exists a unit speed geodesic γ joining x and y, together with a non-trivial Jacobi field along γ that vanishes at x and y. The number of such Jacobi fields that are linearly independent is called the multiplicity of x with respect to y and is always bounded by $n - 1$. When the multiplicity equals $n - 1$ the point x is said to be maximally conjugate to y. As a consequence of our geodesic beam techniques, we obtain quantitative improvements on the L^∞ norm of an eigenfunction near a point x that, loosely speaking, is not maximally self-conjugate.

We work with the set Ξ of unit speed geodesics on (M, g) and for $t \in \mathbb{R}, r > 0$ define

$$\mathcal{C}_x^{r,t} := \left\{\gamma(t) \,\middle|\, \gamma \in \Xi, \, \gamma(0) = x, \, \exists\, n - 1 \text{ conjugate points to } x \text{ in } \gamma(t - r, t + r)\right\}, \tag{2.8}$$

where we count conjugate points with multiplicity. Note that if $r_t \to 0^+$ as $|t| \to \infty$, then saying that $x \in \mathcal{C}_x^{r_t,t}$ for t large indicates that x behaves like a point that is maximally self-conjugate. This is the case for every point on the sphere. The following result applies under

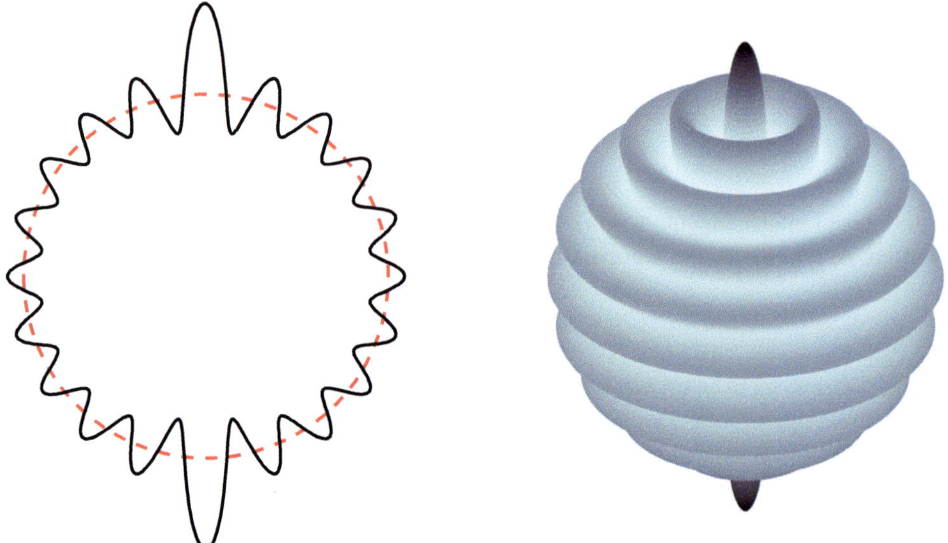

Fig. 2.1 A cross section of a zonal harmonic with the sphere shown in the dashed red line (left) and the graph of a zonal harmonic over the sphere (right)

the assumption that this type of near maximum self-conjugacy does not happen and obtains quantitative improvements in that setting. The obvious case where our next theorem applies is that of manifolds without conjugate points, where $\mathcal{C}_x^{r,t} = \emptyset$ for $0 < r < |t|$. In addition, the theorem applies to *all* non-trivial product manifolds $M = M_1 \times M_2$.

Theorem 2.2.1 ([CG19a, Theorem 1]) *Let $V \subset M$ and assume that there exist $t_0 > 0$ and $a > 0$ so that*

$$\inf_{x \in V} d\big(x, \mathcal{C}_x^{r_t,t}\big) \geq r_t, \qquad for\ t \geq t_0,$$

with $r_t = \frac{1}{a}e^{-at}$. Then, for all $\lambda_0 > 0$ there exist $C > 0$ so that for $\lambda > \lambda_0$ and $u \in \mathcal{D}'(M)$

$$\|u\|_{L^\infty(V)} \leq C\lambda^{\frac{n-1}{2}}\left(\frac{\|u\|_{L^2}}{\sqrt{\log \lambda}} + \lambda\sqrt{\log \lambda}\big\|(-\tfrac{1}{\lambda^2}\Delta_g - 1)u\big\|_{\mathcal{H}_\lambda^{\frac{n-3}{2}}}\right).$$

Here, for $s \in \mathbb{R}$,

$$\|u\|_{\mathcal{H}_\lambda^s}^2 := \langle(-\lambda^{-2}\Delta_g + 1)^s u, u\rangle_{L^2},$$

and we have written $\mathcal{D}'(M)$ for the space of distributions on M.

The theorem from which Theorem 2.2.1 is derived is proved in Sect. 7.1 and is proved using geodesic beam methods. A sketch of the dynamical ideas required to go from Theorem 7.1.1 to Theorem 2.2.1 is given in Chap. 8.

See also Theorem 2.2.12 for a similar statement in which an improved sup-bound on $|u(x)|$ is obtained under a weaker assumption on the volume of the sets of directions generating loops through x.

2.2.2   Averages

One way to measure eigenfunction concentration is to study the average of the eigenfunction over a submanifold $H \subset M$ of codimesion k. Indeed, if u_λ is a Laplace eigenfunction of eigenvalue λ^2, it is known that

$$\int_H u_\lambda d\sigma_H = O(\lambda^{\frac{k-1}{2}}),$$

as $\lambda \to \infty$, where σ_H is the measure on H induce by the Riemannian metric g. This result was proved by Zelditch [Zel92] and is saturated on the round sphere. This generalized the work of Good and Hejhal [Goo83, Hej82]. Chen–Sogge [CS15] were the first to obtain a refinement on the standard bounds. This work has since been improved under various assumptions by Sogge, Xi, Zhang, Wyman, Toth, and the authors [SXZ16, Wym17, Wym20, Wym19, Wym18, CGT18, CG19b].

As in the case of L^∞ norms [CG19b, CG21, CG19a] use geodesic beams to improve on previous estimates. Before stating our next theorem, which illustrates some of these improvements, we recall that if (M, g) has strictly negative sectional curvature, then it also has Anosov geodesic flow [Ano67]. Also, both Anosov geodesic flow [Kli74] and non-positive sectional curvature imply that (M, g) has no conjugate points.

Theorem 2.2.2 ([CG19a, Theorems 3 and 4]) *Let (M, g) be a smooth, compact Riemannian manifold of dimension n. Let $H \subset M$ be a closed embedded submanifold of codimension k. Suppose one of the following assumptions holds:*

A. *(M, g) has no conjugate points and H has codimension $k > \frac{n+1}{2}$.*
B. *(M, g) has no conjugate points and H is a geodesic sphere.*
C. *(M, g) is a surface with Anosov geodesic flow.*
D. *(M, g) is non-positively curved and has Anosov geodesic flow, and H has codimension $k > 1$.*
E. *(M, g) is non-positively curved and has Anosov geodesic flow, and H is totally geodesic.*
F. *(M, g) has Anosov geodesic flow and H is a subset of M that lifts to a horosphere in the universal cover.*

Then, there exists $C > 0$ so that for all $w \in C_c^\infty(H)$ the following holds. There is $\lambda_0 > 0$ so that for $\lambda > \lambda_0$ and $u \in \mathcal{D}'(M)$

$$\left| \int_H w u \, d\sigma_H \right| \le C\lambda^{\frac{k-1}{2}} \|w\|_\infty \left(\frac{\|u\|_{L^2}}{\sqrt{\log \lambda}} + \lambda\sqrt{\log \lambda}\|(-\tfrac{1}{\lambda^2}\Delta_g - 1)u\|_{\mathcal{H}_\lambda^{\frac{k-3}{2}}} \right). \tag{2.9}$$

The bounds in [Wym20, Wym18] are special cases of Theorem **C**, Theorem **D**, and the results of [CG19a, Theorem 6]. We also note that for any smooth compact embedded submanifold, $H_0 \subset M$, satisfying one of the conditions in Theorem 2.2.2, there is a neighborhood U of H_0, in the C^∞ topology, so that the constants C and h_0 in Theorem 2.2.2 are uniform over $H \in U$ and w taken in a bounded subset of $C_c^\infty(H)$. In particular, the sup-norm bounds from [Bér77a, Bon17a, Ran78] are a special case of Theorem **A**. Similar to the $o(h^{\frac{1-k}{2}})$ bounds in [CG19b], we conjecture that (2.9) holds whenever (M, g) is a manifold with Anosov geodesic flow, regardless of the geometry of H.

To the authors' knowledge, the results in [CG19a] improve and extend *all* existing bounds on averages over submanifolds for eigenfunctions of the Laplacian, including those on L^∞ norms (without additional assumptions on the eigenfunctions). Our estimates imply those of [CG19b] and therefore give all previously known improvements of the form $\int_H u\, d\sigma_H = o(h^{\frac{1-k}{2}})$. Moreover, we are able to improve upon the results of [Wym18, Wym20, SXZ16, Bér77a, Bon17a, Ran78].

2.2.3 Weyl Laws

Asymptotics for the spectral projector play a crucial role in the study of eigenvalues and eigenfunctions for the Laplacian, with applications to the study of physical phenomena such as wave propagation and quantum evolution. One of the oldest problems in spectral theory is to understand how the Laplace eigenvalues $\{\lambda_j^2\}_j$ distribute on the real line. Let

$$N(\lambda) := \#\{j : \lambda_j \le \lambda\}$$

be the eigenvalue counting function. Motivated by black body radiation, Hilbert conjectured that, as $\lambda \to \infty$,

$$N(\lambda) = (2\pi)^{-n} \operatorname{vol}_{\mathbb{R}^n}(B) \operatorname{vol}_g(M)\lambda^n + E(\lambda), \tag{2.10}$$

with $E(\lambda) = o(\lambda^{n-1})$. Here, $\operatorname{vol}_{\mathbb{R}^n}(B)$ is the volume of the unit ball $B \subset \mathbb{R}^n$, $\operatorname{vol}_g(M)$ is the Riemannian volume of M, and dv_g is the volume measure induced by the Riemannian metric. The conjecture was proved by Weyl [Wey12] and is known as the Weyl Law. We refer to $E(\lambda)$ as a *Weyl remainder.* In 1968, Hörmander [Hör68], provided a framework for the study of $E(\lambda)$ and generalized the works of Avakumović [Ava56] and Levitan [Lev52], who proved

$$E(\lambda) = O(\lambda^{n-1});$$

a result that is sharp on the round sphere and is thought of as the standard remainder.

The article [Hör68] provided a framework for the study of Weyl remainders which led to many advances, including the work of Duistermaat–Guillemin [DG75] who showed

$E(\lambda) = o(\lambda^{n-1})$ when the set of periodic geodesics has measure 0. In 1977, Bérard [Bér77b] showed that $E(\lambda) = O(\lambda^{n-1}/\log \lambda)$ on both surfaces without conjugate points and non-positively curved manifolds of any dimension. Fifteen years later, Volovoy [Vol90a] provided estimates under dynamical conditions guaranteeing that $E(\lambda) = O(\lambda^{n-1}/\log \lambda)$ and verified these conditions for certain specific examples in [Vol90b]. The recent work of Bonthenneau [Bon17b] improved a geometric estimate in Bérard's work, thus generalizing his result to manifolds without conjugate points of any dimension.

In this section we explain that one can obtain logarithmic improvements on $E(\lambda)$ under appropriate dynamical conditions that generalize the ones introduced above. We will also discuss improved asymptotics for pointwise Weyl Laws. That is, we study asymptotics for the Schwartz kernel of the spectral projector

$$\Pi_\lambda : L^2(M, g) \to \bigoplus_{\lambda_j \leq \lambda} \ker(-\Delta_g - \lambda_j^2),$$

i.e. Π_λ is the orthogonal projection operator onto functions with frequency at most λ. If $\{u_{\lambda_j}\}_{j=1}^{\infty}$ is an orthonormal basis of eigenfunctions, $-\Delta_g u_{\lambda_j} = \lambda_j^2 u_{\lambda_j}$, the Schwartz kernel of Π_λ is

$$\Pi_\lambda(x, y) = \sum_{\lambda_j \leq \lambda} u_{\lambda_j}(x)\overline{u_{\lambda_j}}(y), \qquad (x, y) \in M \times M.$$

Asymptotics for the spectral projector play a crucial role in the study of eigenvalues and eigenfunctions for the Laplacian, with applications to the study of physical phenomena such as wave propagation and quantum evolution.

2.2.3.1 On Diagonal Estimates

We next discuss $E_\lambda(x)$, the remainder in the on diagonal pointwise Weyl law

$$\Pi_\lambda(x, x) = (2\pi)^{-n} \operatorname{vol}_{\mathbb{R}^n}(B)\lambda^n + E_\lambda(x), \qquad x \in M. \tag{2.11}$$

It is known that [Hör68]

$$E_\lambda(x) = O(\lambda^{n-1}), \qquad x \in M,$$

and that the bound is, again, sharp on the round sphere. Note that

$$E(\lambda) = \int_M E_\lambda(x) \, \mathrm{dv}_g(x).$$

The connection between $E_\lambda(x)$ and geodesic loops through x is studied in the works of Safarov, Sogge–Zelditch [Saf88, SZ02b] and often appears in estimates for sup-norms of eigenfunctions.

The connection between the spectrum of the Laplacian and the properties of periodic geodesics on M has been known since at least the works [Cha74, CdV73, Wei75], with their relation to Weyl remainders first explored in the seminal work [DG75]. To control $E(\lambda)$

we impose dynamical conditions on the periodicity properties of the geodesic flow $\varphi_t :$ $T^*M \setminus \{0\} \to T^*M \setminus \{0\}$, i.e., the Hamiltonian flow of $(x, \xi) \mapsto |\xi|_{g(x)}$. For $t_0 > 0$, $T > 0$, and $R > 0$, define the set of near periodic directions in $U \subset S^*M$ by

$$\mathcal{P}_U^R(t_0, T) := \left\{ \rho \in U : \bigcup_{t_0 \leq |t| \leq T} \varphi_t(B_{S^*M}(\rho, R)) \cap B_{S^*M}(\rho, R) \neq \emptyset \right\}. \tag{2.12}$$

Given two sets $U \subset V \subset T^*M$, and $R > 0$, we write $B_V(U, R) := \{\rho \in V : d(U, \rho) < R\}$, where d is the distance induced by some fixed metric on T^*M, $B(U, R) = B_{T^*M}(U, R)$, and $B_V(\rho, R) = B_V(\{\rho\}, R)$.

We phrase our dynamical conditions in terms of a resolution function $\mathbf{T} = \mathbf{T}(R)$. This is a function of the scale, R, at which the manifold is resolved, which increases as $R \to 0^+$. We use $\mathbf{T}$ to measure the time for which balls of radius R can be propagated under the geodesic flow while satisfying a given dynamical assumption, e.g. being non periodic.

Definition 2.2.3 (*Resolution function*) We say a decreasing, continuous function $\mathbf{T} : (0, \infty)$ $\to (0, \infty)$ is a *resolution function*. In addition, we say a resolution function $\mathbf{T}$ is *sub-logarithmic*, if it is differentiable and

$$(\log \log R^{-1})' = -1 \big/ R \log R^{-1} \leq [\log \mathbf{T}(R)]' \leq 0, \qquad 0 < R < 1.$$

We measure how close $\mathbf{T}$ is to being logarithmic through

$$\Omega(\mathbf{T}) := \limsup_{R \to 0^+} \mathbf{T}(R) \big/ \log R^{-1}. \tag{2.13}$$

Simple examples of sub-logarithmic resolution functions are $\mathbf{T}(R) = \alpha(\log R^{-1})^\beta$ for any $\alpha > 0$ and $0 < \beta \leq 1$.

For improved integrated Weyl remainders, we need a condition on the geodesic flow.

Definition 2.2.4 (**T** *non-periodic*) Let $\mathbf{T}$ be a resolution function. Then $U \subset S^*M$ is said to be $\mathbf{T}$ *non-periodic* with constant C_{np} provided there exists $t_0 > 0$ such that

$$\limsup_{R \to 0^+} \mu_{S^*M}\left(B_{S^*M}\left(\mathcal{P}_U^R(t_0, \mathbf{T}(R)), R \right) \right) \mathbf{T}(R) \leq C_{\text{np}}.$$

We say U is $\mathbf{T}$ non-periodic if there is such C_{np}, and $W \subset M$ is $\mathbf{T}$ non-periodic if S_W^*M is.

Next, we present our main theorem, which obtains improvements for $E(\lambda)$ and $E_\lambda(x)$.

Theorem 2.2.5 ([CG20b, Theorem 2]) *Let (M, g) be a smooth compact connected Riemannian manifold of dimension n, $W \subset M$ be an open subset with $\dim_{\text{box}} \partial W < n$, and $\Omega_0 > 0$.*

Table 2.1 The table list examples where the assumptions of Theorem 2.2.5 hold. The table lists examples with **T** non-periodic subsets with $\mathbf{T}(R) = c \log R^{-1}$. The final two examples are due to Volovoy [Vol90b]. In the special case $W = M$, Theorem 2.2.5 gives a different proof of the weaker statement in [Vol90a, Theorem 0.1]

M	W	§
Product manifolds	Any	[CG20b, §B.1.1]
Perturbed spheres	In the non-periodic set	[CG20b, §B.2.1]
Manifolds without conjugate points	Any	[CG20b, §B.1]
Non-Zoll convex analytic surfaces of revolution	Any	[Vol90b]
Compact Lie group rank > 1 with bi-invariant metric	Any	[Vol90b]

There exists $C_0 > 0$ such that if $\mathbf{T}$ is a sub-logarithmic rate function with $\Omega(\mathbf{T}) < \Omega_0$ and W is $\mathbf{T}$ non-periodic, then there is $\lambda_0 > 0$ such that for all $\lambda > \lambda_0$

$$\left| \int_W E_\lambda(x) \, \mathrm{d}v_g(x) \right| \leq C_0 \, \lambda^{n-1} \big/ \mathbf{T}\big(\lambda^{-1}\big).$$

In particular, if M is $\mathbf{T}$ non-periodic, then there is λ_0 such that for all $\lambda > \lambda_0$

$$|E(\lambda)| \leq C_0 \, \lambda^{n-1} \big/ \mathbf{T}\big(\lambda^{-1}\big).$$

For some example applications of Theorem 2.2.5, we refer the reader to Table 2.1. We prove Theorem 2.2.5 in the case $W = M$ in Sect. 2.2.3. After reformulating the counting function as the integral of the kernel of the spectral projector over the diagonal and identifying this kernel with a quasimode, the proof applies geodesic beam estimates of averages of quasimodes to control the remainder.

Theorem 2.2.5 fits in a long history of work on asymptotics of the kernel of the spectral projector and the eigenvalue counting function. Many authors considered pointwise Weyl sums [MP49, Gr53, Ava56, Lev52, See67, Hör68], eventually proving the sharp remainder estimates. The article [Hör68] provided a method which was used in many later works: [DG75] showed $E(\lambda) = o(\lambda^{n-1})$ under the assumption that the set of periodic trajectories has measure 0, [Saf88, SZ02b] improved estimates on $E_\lambda(x)$ to $o(\lambda^{n-1})$ under the assumption that the set of looping directions through x has measure 0 (see also the book of Safarov–Vassiliev [SV97]).

2.2.3.2 Off Diagonal Estimates

In this section, we present improved estimates for the off-diagonal $\Pi_\lambda(x, y)$. To do that, we first introduce the relevant dynamical conditions.

For $t_0 > 0$, $T > 0$, $R > 0$, and $x, y \in M$, define

$$\mathcal{L}^R_{x,y}(t_0, T) := \left\{ \rho \in S^*_x M : \bigcup_{t_0 \le |t| \le T} \varphi_t(B(\rho, R)) \cap B(S^*_y M, R) \neq \emptyset \right\}. \tag{2.14}$$

Definition 2.2.6 (*Non looping pair*) Let $\mathbf{T}$ be a resolution function, $t_0 > 0$, $C_{\mathrm{nl}} > 0$, and $x, y \in M$. Then, (x, y) *is said to be a* $(t_0, \mathbf{T})$ *non-looping pair with constant* C_{nl} *when*

$$\limsup_{R \to 0^+} \left(\mu_{S^*_x M} \left(B_{S^*_x M}(\mathcal{L}^R_{x,y}(t_0, \mathbf{T}(R)), R) \right) \mu_{S^*_y M} \left(B_{S^*_y M}(\mathcal{L}^R_{y,x}(t_0, \mathbf{T}(R)), R) \right) \mathbf{T}(R)^2 \right) \le C_{\mathrm{nl}}.$$

We say x is $(t_0, \mathbf{T})$ *non-looping with constant* C_{nl} if (x, x) is $(t_0, \mathbf{T})$ non-looping with constant C_{nl}.

Note that if $t_0 < \mathrm{inj}(M)$, where $\mathrm{inj}(M)$ is the injectivity radius of M, then for x to be $(t_0, \mathbf{T})$ non-looping is the same as being $(\varepsilon, \mathbf{T})$ non-looping for any $0 < \varepsilon \le t_0$. In this case, we write x is $(0, \mathbf{T})$ non-looping.

If (x, y) is a $(t_0, \mathbf{T})$ non looping pair for some $t_0 > 0$ we measure the difference between $\Pi_\lambda(x, y)$ and its smoothed version which takes into account propagation up to time t_0. Let $\rho \in \mathscr{S}(\mathbb{R})$ with $\hat\rho(0) \equiv 1$ on $[-1, 1]$ and $\mathrm{supp}\,\hat\rho \subset [-2, 2]$. For $\sigma > 0$ we define

$$\rho_\sigma(s) := \sigma \rho(\sigma s). \tag{2.15}$$

For $x, y \in M$, $t_0 > 0$, and $\lambda > 0$, let

$$E^{t_0}_\lambda := \Pi_\lambda - \rho_{t_0} * \Pi_\lambda, \tag{2.16}$$

where the convolution is taken in the λ variable.

The choice of $E^{t_0}_\lambda$ as the relevant error may seem strange. However, the term $\rho_{t_0} * \Pi_\lambda$ can be written using only a short-time propagator for the wave equation and therefore can be understood accurately using the classical theory of Fourier integral operators developed by Hörmander. Therefore, understanding the error in (2.16) well is equivalent to understanding Π_λ well.

Theorem 2.2.7 ([CG20b, Theorem 4]) *Let* $\alpha, \beta \in \mathbb{N}^n$, $0 < \delta < \frac{1}{2}$, $C_{\mathrm{nl}} > 0$, $R_0 > 0$, $\Omega_0 > 0$, $\varepsilon > 0$, *and* t *be a resolution function, there is* $C_0 > 0$ *such that if* $\mathbf{T}_j$ *is a sub-logarithmic resolution function with* $\Omega(\mathbf{T}_j) < \Omega_0$ *for* $j = 1, 2$ *and* $\mathbf{T}_{\max} = \max(\mathbf{T}_1, \mathbf{T}_2)$, *then there is* $\lambda_0 > 0$ *such the following holds. If* $x_0, y_0 \in M$ *and* $t_0 > 0$ *are such that* x_0 *and* y_0 *are*

Table 2.2 The table lists examples where Theorem 2.2.7 holds. We write (nL) when H_i is $\mathbf{T}_i$ non-looping and (nRvc) when H_i is $\mathbf{T}_i$ non-recurrent via coverings

	x	y	$\mathbf{T}_1$	$\mathbf{T}_2$	§
Product manifolds	Any point (nL)	Any point (nL)	$\log R$	$\log R$	[CG20b, §B.1.1]
Spherical pendulum	Not a pole (nL)	$y = x(nL)$	$\log R$	$\log R$	[CG20b, §B.2.2]
Spherical pendulum	Not a pole (nL)	A pole	$\log R$	$\mathrm{inj}\, M$	[CG20b, §B.2.2]
Perturbed spheres	Non-periodic, not a pole, (nL)	A pole	$\log R$	$\mathrm{inj}\, M$	[CG20b, §B.2.1]
Perturbed spheres	Non-periodic, not a pole, (nL)	$y = x\ (nL)$	$\log R$	$\log R$	[CG20b, §B.2.1]

respectively $(\mathfrak{t}, \mathbf{T}_1)$ *and* $(\mathfrak{t}, \mathbf{T}_2)$ *non-recurrent at scale* R_0, *and* (x_0, y_0) *is a* $(t_0, \mathbf{T}_{\max})$ *non-looping pair with constant* C_{nl}, *then for* $\lambda > \lambda_0$

$$\sup_{x \in B(x_0, \lambda^{-\delta})} \sup_{y \in B(y_0, \lambda^{-\delta})} \left| \partial_x^\alpha \partial_y^\beta E_\lambda^{t_0 + \varepsilon}(x, y) \right| \leq C_0\, \lambda^{n-1+|\alpha|+|\beta|} \Big/ \sqrt{\mathbf{T}_1(\lambda^{-1})\mathbf{T}_2(\lambda^{-1})}.$$

In Table 2.2 we list some examples for which Theorem 2.2.7 holds. In each case there exists $t_0 > 0$ such that (x, y) is a $(t_0, \max(\mathbf{T}_1, \mathbf{T}_2))$ non-looping pair. Note that we omit labeling points for which $\mathbf{T}_2 = \mathrm{inj}(M)$ since being $\mathrm{inj}(M)$ non-recurrent is an empty statement. In these cases the gain in the pointwise Weyl law is $\sqrt{\log \lambda}$ instead of $\log \lambda$.

The proof of Theorem 2.2.7 (or more precisely a sketch of the changes necessary from the proof of Theorem 2.2.5) is given in Sect. 7.4.7. As with Theorem 2.2.5, we are able to convert questions about the error into ones about pointwise estimates on quasimodes at which point we may apply the geodesic beam estimates developed in Chap. 6.

2.2.3.3 Logarithmic Improvements on the Weyl Law are Typical

In this section we explain that given a manifold M the bound on $E(\lambda)$ proved in Theorem 2.2.5 (with $\mathbf{T}(\lambda) = (\log \lambda)^{1/\alpha}$ for some $\alpha > 1$) holds for most Riemmanian metrics.

Since the space of Riemannian metrics, $\mathscr{G}$, on a manifold cannot be endowed with a non-trivial, translation invariant Borel measure, we introduce an analog of full Lebesgue measure in infinite dimensions called *predominance*. We then study properties of the geodesic flow and remainders in the Weyl law for predominant sets of metrics.

The notion of predominance has three important properties: (1) any predominant set is dense, (2) a finite intersection of predominant sets is predominant, and (3) in finite dimensions, a predominant set has full Lebesgue measure. Heuristically, a set $G \subset \mathscr{G}$ is *predomi-*

nant if there is a family of submanifolds endowed with finite Borel measures $\{(\Gamma_g, \mu_g)\}_{g \in \mathcal{G}}$ such that $g \in \Gamma_g$, μ_g assigns a positive measure to any neighborhood of g, the map $g \mapsto \Gamma_g$ is C^1, and $G \cap \Gamma_g$ has full μ_g measure for every $g \in \mathcal{G}$. For the careful definition of this concept, see [CG22, Definition 2.4].

Since we aim to study how the remainder in the Weyl Law behaves for most metrics g, we proceed to keep track of the metric dependence and write $E(\lambda, g)$ in place of $E(\lambda)$ (as defined in (2.10)). Our main result shows that the Weyl law has a *logarithmic* improvement for a predominant set of metrics.

Theorem 2.2.8 ([CG22, Theorem 1.1]) *Let $d \geq 2$. There is $\nu_0 > 0$ and for all $\nu > 0$ there is $\Omega_\nu > 0$ such that the following holds. If $\nu \geq \nu_0$, M is a compact C^ν-manifold of dimension n without boundary, and $\Omega > \Omega_\nu$, then there is a predominant set $G_\Omega \subset \mathcal{G}^\nu$ such that for every $g \in G_\Omega$*

$$E(\lambda, g) = O_g\big(\lambda^{n-1}\big/(\log \lambda)^{\frac{1}{\Omega}}\big),$$

as $\lambda \to \infty$. In particular, G_Ω is dense in $\mathcal{G}^\nu$.

The constant Ω_ν in Theorem 2.2.8 is explicit, and we can take $\Omega_\nu := 1 + \log_2(2(d-1)(2d+1)(\max(\nu, 6) + 3d - 1)) - \log_2(2d - 1)$.

As explained before, for manifolds with no conjugate points the works of Bérard [Bér77b] and Bonthenneau [Bon17b] yields that $E(\lambda, g) = O_g(\lambda^{n-1}/\log \lambda)$. In addition, Volovoy [Vol90a] provided estimates under dynamical conditions guaranteeing that $E(\lambda, g) = O_g(\lambda^{n-1}/\log \lambda)$ and verified these conditions for certain specific examples in [Vol90b]. Manifolds where there are known polynomial improvements of the form $E(\lambda, g) = O_g(\lambda^{n-1-\varepsilon})$ are very rare. For instance, such estimates hold on the torus [Hux03, BW17], products of spheres [IW21], and other special integrable systems [Vol90b]. Nevertheless, it has long been expected that, for a 'typical' metric g, there exists $\varepsilon > 0$ such that $E(\lambda, g) = O_g(\lambda^{n-1-\varepsilon})$. However, until now, the best available result is that $E(\lambda, g) = o(\lambda^{d-1})$ for a Baire-generic set of g. This can be recovered from the work of Sogge–Zelditch [SZ02b] or can be seen by combining the remainder estimates in [DG75] with the bumpy metric theorem of Anosov and Abraham [Ano82, Abr70].

Theorem 2.2.8 improves on these bounds in two important ways. First, Baire genericity is replaced by the concept of predominance which is an analog of full Lebesgue measure in infinite dimensions. Just as in finite dimensions a full Lebesgue measure set is much more 'typical' than a Baire generic one (indeed, a Baire generic set may have measure 0), a predominant set in infinite dimensions is much more 'typical' than a Baire generic set. Second, although the change from $o(\lambda^{n-1})$ to $O(\lambda^{n-1}/(\log \lambda)^{1/\Omega})$ may seem small, this improvement requires the development of new ideas and requires subtle dynamical estimates. In addition, it is the only quantitative remainder estimate available for typical metrics.

Theorem 2.2.8 is proved by combining the result in Theorem 2.2.5 with a new result, [CG22, Theorem 1.3], that proves that (M, g) is **T**-non periodic for a predominat set of metrics g, with

$$\mathbf{T}(R) = \log(R^{-1})^{-1/\Omega}.$$

The proof of [CG22, Theorem 1.3] is purely dynamical in nature, so we do not include it here.

2.2.4 L^p Norms

Since the work of Sogge [Sog88], it has been known that there is a change of behavior in the growth of L^p norms for eigenfunctions at the *critical exponent* $p_c := \frac{2(n+1)}{n-1}$. If u_λ is a Laplace eigenfunction of eigenvalue λ^2, Sogge proved that

$$\|u_\lambda\|_{L^p} \leq C\lambda^{\delta(p)}, \qquad \delta(p) := \begin{cases} \frac{n-1}{2} - \frac{n}{p} & p_c \leq p \\ \frac{n-1}{4} - \frac{n-1}{2p} & 2 \leq p \leq p_c. \end{cases} \tag{2.17}$$

For $p \geq p_c$, (2.17) is saturated by the zonal harmonics on the round sphere, S^n. See Fig. 2.1. On the other hand, for $p \leq p_c$, these bounds are saturated by the highest weight spherical harmonics on S^n, also known as Gaussian beams. See Fig. 2.2.

In Theorem 2.2.10 below, we show, in a very strong sense, how any eigenfunction saturating (2.17) for $p > p_c$ behaves like a zonal harmonic. At the same time, Blair–Sogge [BS15a, BS17a] showed that for $p < p_c$ such eigenfunctions behave like Gaussian beams. In the

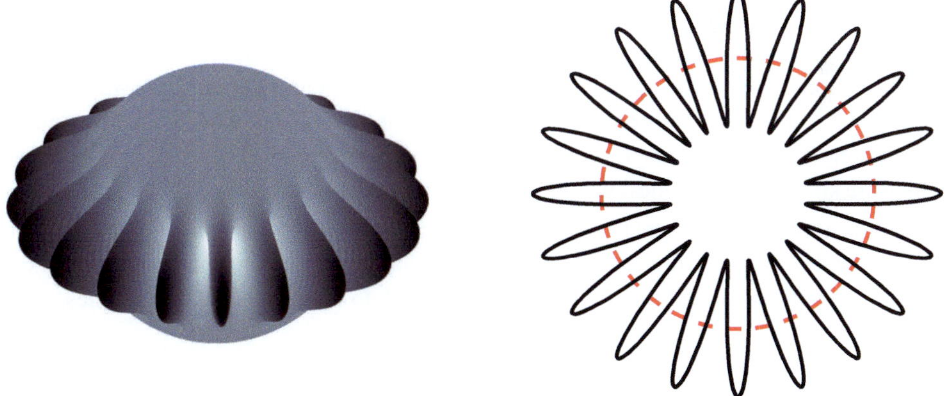

Fig. 2.2 The highest weight spherical harmonic plotted as a graph over the sphere (Left). The cross section of the graph along the equator. (Right)

$p \leq p_c$, Blair–Sogge have recently made substantial progress on improved L^p estimates on manifolds with non-positive curvature [BS19, BS18, BS15b, BHS22].

When (M, g) has non-positive sectional curvature, Hassell and Tacy [HT15] gave quantitative gains over this estimate of the form

$$\|u_\lambda\|_{L^p} = O(\lambda^{\delta(p)}/(\log \lambda)^{\sigma(p)})$$

when $p > p_c$ and with $\sigma(p) = \frac{1}{2}$. Blair and Sogge [BS17b, Bla10a] also obtained an improvement when $2 < p \leq p_c$ for some $\sigma(p) > 0$ smaller than $\frac{1}{2}$.

Next, we present applications of our geodesic beam techniques which yield $\sqrt{\log \lambda}$ improvements for L^p norms with $p > p_c$, generalizing those of [HT15]. We continue to work with $\mathcal{C}_x^{r,t}$ as defined in (2.8). Note that if $r_t \to 0^+$ as $|t| \to \infty$, then saying $y \in \mathcal{C}_x^{r_t,t}$ for t large indicates that y behaves like a point that is maximally conjugate to x. This is the case for every point x on the sphere when y is either equal to x or its antipodal point. The following result applies under the assumption that points are not maximally conjugate and obtains quantitative improvements.

Theorem 2.2.9 ([CG20a, Theorem 1]) *Let $p > p_c$, $U \subset M$, and assume there exist $t_0 > 0$ and $a > 0$ so*

$$\inf_{x_1,x_2 \in U} d\left(x_1, \mathcal{C}_{x_2}^{r_t,t}\right) \geq r_t, \qquad \textit{for } t \geq t_0,$$

with $r_t = \frac{1}{a}e^{-at}$. Then, there exist $C > 0$ and $\lambda_0 > 0$ so that for $\lambda > \lambda_0$ and $u \in \mathcal{D}'(M)$

$$\|u\|_{L^p(U)} \leq C\lambda^{\delta(p)} \left(\frac{\|u\|_{L^2}}{\sqrt{\log \lambda}} + \lambda\sqrt{\log \lambda}\left\|(-\tfrac{1}{\lambda^2}\Delta_g - 1)u\right\|_{\mathcal{H}_\lambda^{\frac{n-3}{2} - \frac{n}{p}}} \right).$$

The proof of Theorem 2.2.9 gives a great deal of information about eigenfunctions which may saturate L^p bounds ($p > p_c$). Our next theorem describes the structure of such eigenfunctions. This theorem shows that an eigenfunction can saturate the *logarithmically improved L^p* norm near at most *boundedly many* points. Moreover, modulo an error small in L^p, near each of these points the eigenfunction can be decomposed as a sum of quasimodes which are similar to the highest weight spherical harmonics scaled by $\lambda^{-\frac{n-1}{4}}/\sqrt{\log \lambda}$ whose number is nearly proportional to $\lambda^{\frac{n-1}{2}}$. In the theorem below the quasimodes are denoted by v_j and, while similar to highest weight spherical harmonics (a.k.a Gaussian beams), they are not as tightly localized to a geodesic segment and do not have Gaussian profiles.

Theorem 2.2.10 ([CG20a, Theorem 2]) *Let $p > p_c$. There exist $c, C > 0$ such that the following holds. Suppose the same assumptions as Theorem 2.2.9. Let $0 < \delta_1 < \delta_2 < \frac{1}{2}$, $\lambda^{-\delta_2} \leq R(\lambda) \leq \lambda^{-\delta_1}$, and $\{x_\alpha\}_{\alpha \in \mathcal{I}(\lambda)} \subset M$ be a maximal $R(\lambda)$-separated set. Let $u \in \mathcal{D}'(M)$ with $\|(-\frac{1}{\lambda^2}\Delta_g - 1)u\|_{H_h^{\frac{n-3}{2}}} = o\left(\frac{1}{\lambda \log \lambda}\|u\|_{L^2}\right)$, and for $\varepsilon > 0$ let*

$$\mathcal{S}_U(\lambda, \varepsilon, u) := \left\{ \alpha \in \mathcal{I}(\lambda) : \|u\|_{L^\infty(B(x_\alpha, R(\lambda)))} \geq \frac{\varepsilon \lambda^{\frac{n-1}{2}}}{\sqrt{\log \lambda}} \|u\|_{L^2}, \quad B(x_\alpha, R(\lambda)) \cap U \neq \emptyset \right\}.$$

Then, for all $\varepsilon > 0$ there are $N_\varepsilon > 0$ and $\lambda_0 > 0$ such that $|\mathcal{S}_U(\lambda, \varepsilon, u)| \leq N_\varepsilon$ for all $\lambda > \lambda_0$.

Moreover, there is collection of geodesic tubes $\{\mathcal{T}_j\}_{j \in \mathcal{L}(\varepsilon, u)}$ of radius $R(\lambda)$ (see Definition 6.2.3), with indices satisfying $\mathcal{L}(\varepsilon, u) = \cup_{i=1}^C \mathcal{J}_i$ and $\mathcal{T}_k \cap \mathcal{T}_\ell = \emptyset$ for $k, \ell \in \mathcal{J}_i$ with $k \neq \ell$, such that

$$u = u_e + \frac{1}{\sqrt{\log \lambda}} \sum_{j \in \mathcal{L}(\varepsilon, u)} v_j,$$

where v_j is microsupported in $\mathcal{T}_j$, $|\mathcal{L}(\varepsilon, u)| \leq C \varepsilon^{-2} R(\lambda)^{1-n}$, and for all $p \leq q \leq \infty$,

$$\|u_e\|_{L^q} \leq \varepsilon \lambda^{\delta(q)} (\log \lambda)^{-\frac{1}{2}} \|u\|_{L^2},$$

$$\|v_j\|_{L^2} \leq C \varepsilon^{-1} R(\lambda)^{\frac{n-1}{2}} \|u\|_{L^2}, \qquad \|Pv_j\|_{L^2} \leq C \varepsilon^{-1} R(\lambda)^{\frac{n-1}{2}} \lambda^{-1} \|u\|_{L^2}.$$

Finally, with $\mathcal{L}(\varepsilon, u, \alpha) := \left\{ j \in \mathcal{L}(\varepsilon, u) : \pi(\mathcal{T}_j) \cap B(x_\alpha, 3R(\lambda)) \neq \emptyset \right\}$, for every $\alpha \in \mathcal{S}_U$ $(\lambda, \varepsilon, u)$,

$$c \varepsilon^2 R(\lambda)^{1-n} \leq |\mathcal{L}(\varepsilon, u, \alpha)| \leq C R(\lambda)^{1-n}, \qquad \sum_{j \in \mathcal{L}(\varepsilon, u, \alpha)} \|v_j\|_{L^2}^2 \geq c^2 \varepsilon^2.$$

The notion of the microsupport of a function is defined in Definition 3.2.11.

We do not prove Theorem 2.2.10 in this book, however, it can be derived from the proof of Theorem 7.3.1 below.

The decomposition of u into geodesic beams v_j is illustrated in Fig. 2.3. One covers S^*M with a collection of tubes $\{\mathcal{T}_j\}$ of radius $R(\lambda)$ that run along a geodesic. Each geodesic beam v_j corresponds to microlocalizing u to the tube $\mathcal{T}_j$.

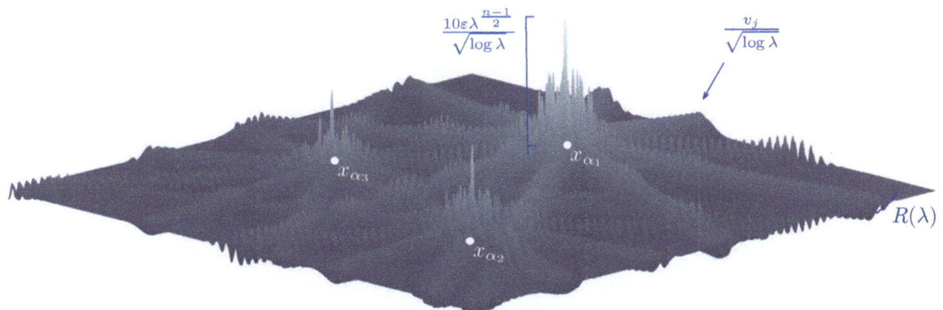

Fig. 2.3 The figure illustrates a function u that saturates the L^∞ bound at three points $x_{\alpha_1}, x_{\alpha_2}, x_{\alpha_3}$ viewed as a superposition of geodesic beams v_j. Each ridge corresponds to a beam v_j and is microsupported on a tube $\mathcal{T}_j$ of radius $R(\lambda)$

Let u be a quasimode with $\|Pu\| = o(1/(\lambda \log \lambda))\|u\|$. Note that, by interpolation Theorem 2.2.9 implies that for each $p > p_c$ there is $N > 0$ such that if $\|u\|_{L^p} \geq \varepsilon \lambda^{\delta(p)}/\sqrt{\log \lambda}\|u\|_{L^2}$, then $\|u\|_{L^\infty} \geq \varepsilon^N \lambda^{\frac{n-1}{2}}\|u\|_{L^2}$. In particular, for u to saturate the logarithmically improved L^p bound, it follows that $\mathcal{S}_M(\lambda, \varepsilon^N, u)$ is non-empty. Theorem 2.2.10 then gives that $\mathcal{S}_M(\lambda, \varepsilon^N, u)$ has a uniformly bounded number of points and at these points the quasimode u needs to consist of at least $c\varepsilon^{2N} R(\lambda)^{1-n}$ geodesic beams whose combined L^2 mass is at least $c\varepsilon^N$. Since $\dim(S^*_{x_\alpha} M) = n - 1$, this implies that there is a positive measure set of directions through x_α among which u is spreading its mass nearly uniformly.

Theorem 2.2.11 *For all $p > p_c$, C_{nl}, $t_0 > 0$, $\Omega_0 > 0$, $\varepsilon_0 > 0$, there is $C > 0$ such that if $U \subset M$, $\mathbf{T}$ is a sub-logarithmic resolution function with $\Omega(\mathbf{T}) < \Omega_0$, and*

$$\sup_{x,y \in U} \limsup_{R \to 0^+} \mu_{S^*_x M}\left(B_{S^*_x M}\left(\mathcal{L}^R_{x,y}(t_0, \mathbf{T}(R)), R\right)\right)\mathbf{T}(R)^{1 - \frac{p_c}{p}} \, {}^{\frac{3+\varepsilon_0}{}} \leq C_{\mathrm{nl}},$$

then there is $\lambda_0 > 0$ such that for all $u \in \mathcal{D}'(M)$, and $\lambda > \lambda_0$,

$$\|u\|_{L^p(U)} \leq C \frac{\lambda^{\delta(p)}}{\sqrt{\mathbf{T}(\lambda)}}\left(\|u\|_{L^2} + \lambda \mathbf{T}(\lambda)\|(-\lambda^{-2}\Delta_g - 1)u\|_{\mathcal{H}_\lambda^{\frac{n-3}{2} - \frac{n}{p}}}\right).$$

We sketch the proof of Theorem 2.2.11 using geodesic beam methods in Sect. 2.2.4 (see Theorem 7.3.1). A sketch of the dynamical ideas required to go from Theorem 7.3.1 to Theorems 2.2.11 and 2.2.9 is given in Chap. 8.

In fact, for $p = \infty$, we have the following theorem.

Theorem 2.2.12 *For all C_{nl}, $t_0 > 0$, $\Omega_0 > 0$, there is $C > 0$ such that any $x \in M$ if $\mathbf{T}$ is a sub-logarithmic resolution function with $\Omega(\mathbf{T}) < \Omega_0$, and*

$$\limsup_{R \to 0^+} \mu_{S^*_x M}\left(B_{S^*_x M}\left(\mathcal{L}^R_{x,x}(t_0, \mathbf{T}(R)), R\right)\right)\mathbf{T}(R) \leq C_{\mathrm{nl}},$$

then there is $\lambda_0 > 0$ such that for all $u \in \mathcal{D}'(M)$, and $\lambda > \lambda_0$,

$$|u(x)| \leq C \frac{\lambda^{\frac{n-1}{2}}}{\sqrt{\mathbf{T}(\lambda)}}\left(\|u\|_{L^2} + \lambda \mathbf{T}(\lambda)\|(-\lambda^{-2}\Delta_g - 1)u\|_{\mathcal{H}_\lambda^{\frac{n-3}{2}}}\right).$$

This theorem is a consequence of our main result for L^∞ estimate (Theorem 7.1.5). We sketch the dynamical analysis necessary to obtain it from Theorem 7.1.5 in Sect. 8.1.

Throughout this text we use semiclassical pseudodifferential operators to understand Laplace eigenfunctions and quasimodes. This calculus of pseudodifferential operators is effective for studying asymptotics of solutions to PDE with a small parameter, h. In our case, $h := \lambda^{-1}$, where λ^2 is the Laplace eigenvalue. Semiclassical pseudodifferential operators will allow us to decompose quasimodes into pieces that are localized in position-momentum (phase) space so that we may study each piece using tools adapted to the nature of the PDE at that point in phase space.

3.1   Basic Definitions: Pseudodifferential Operators

3.1.1   Pseudodifferential Operators on $\mathbb{R}^n$

Pseudodifferential operators are quantizations of observables on the phase space, $T^*\mathbb{R}^n$, i.e. of functions $a = a(x, \xi) \in C^\infty(T^*\mathbb{R}^n)$ where we use coordinates (x, ξ) with $x \in \mathbb{R}^n$ and $\xi \in T^*_x\mathbb{R}^n$. In order to define pseudodifferential operators carefully, we first need to define symbol classes. In what follows, given $(x, \xi) \in T^*\mathbb{R}^n$ we write $\langle \xi \rangle := (1 + |\xi|^2)^{1/2}$.

Remark 3.1.1 In order to simplify notation, we typically allow functions and operators to implicitly depend on the small parameter h.

Definition 3.1.2 (*Symbol class*) We say that $a \in C^\infty(T^*\mathbb{R}^n)$ is a symbol of order $m \in \mathbb{R}$ and class $0 \le \delta < \frac{1}{2}$ if for all $\alpha, \beta \in \mathbb{N}^n$, there is $C_{\alpha\beta} > 0$ such that

© The Author(s), under exclusive license to Springer Nature Switzerland AG 2023   25
Y. Canzani and J. Galkowski, *Geodesic Beams in Eigenfunction Analysis*, Synthesis
Lectures on Mathematics & Statistics, https://doi.org/10.1007/978-3-031-31586-2_3

$$|\partial_x^\alpha \partial_\xi^\beta a(x,\xi)| \leq C_{\alpha\beta} h^{-\delta(|\alpha|+|\beta|)} \langle \xi \rangle^{m-|\beta|}.$$

In this case, we write $a \in S_\delta^m(T^*\mathbb{R}^n)$.

We will often write simply $a \in S^m$ when the space is clear from context. We also define $S^\infty := \bigcup_m S^m$, $S^{-\infty} := \bigcap_m S^m$. We also define S^{comp} to be the set of $a \in S^{-\infty}$ which are supported in some h-independent compact set. Furthermore, we often write $S_0^m = S^m$, i.e., omit the $\delta = 0$ in various spaces of symbol classes below.

It will also be be convenient to have a notion of semiclassical Sobolev spaces. In order to define these spaces, we first recall some standard definitions.

Definition 3.1.3 (*Shwartz functions and distributions*) We define the space of Schwartz functions on $\mathbb{R}^n$ by

$$\mathscr{S}(\mathbb{R}^n) := \{u \in C^\infty(\mathbb{R}^n) \ : \ \sup_x |x|^\beta |\partial_x^\alpha u(x)| < C_{\alpha\beta}, \ \text{ for all } \alpha, \beta \in \mathbb{N}^n\}.$$

The space of Schwartz distributions, $\mathscr{S}'(\mathbb{R}^n)$ is then the dual of $\mathscr{S}(\mathbb{R}^n)$.

We next recall the semiclassical Fourier transform.

Definition 3.1.4 (*Semiclassical Fourier transform*) The semiclassical Fourier transform is the map $\mathcal{F} : \mathscr{S}'(\mathbb{R}^n) \to \mathscr{S}'(\mathbb{R}^n)$ given by

$$\mathcal{F}(u)(\xi) := \int_{\mathbb{R}^n} e^{\frac{i}{h}\langle x-y,\xi \rangle} u(y)\,dy.$$

We can now define the semiclassical Sobolev spaces. The elements of these spaces are the same as for the standard Sobolev spaces, but the norm is scaled in a way depending on h.

Definition 3.1.5 (*Semiclassical Sobolev norm*) For $s \in \mathbb{R}$ the s-semiclassical Sobolev norm is defined as

$$H_h^s(\mathbb{R}^n) := \{u \in \mathscr{S}'(\mathbb{R}^n) \ : \ \langle \xi \rangle^s \mathcal{F}(u) \in L^2(\mathbb{R}^n)\}, \qquad \|u\|_{H_h^s}^2 := (2\pi h)^{-n} \|\langle \xi \rangle^s \mathcal{F}(u)\|_{L^2}^2.$$

For $A : \mathscr{S}(\mathbb{R}^n) \to \mathscr{S}(\mathbb{R}^n)$, we say that $A = O(h^\infty)_{\psi-\infty}$ if for all N, there is $C_N > 0$ such that

$$\|A\|_{H_h^{-N} \to H_h^N} \leq C_N h^N.$$

We may now introduce the class of pseudodifferential operators on $\mathbb{R}^n$.

Definition 3.1.6 (*Pseudodifferential operator on* $\mathbb{R}^n$) For $m \in \mathbb{R}$, we say that A is a *pseudodifferential operator of order m* and write $A \in \Psi_\delta^m(\mathbb{R}^n)$ if there is $a \in S_\delta^m(T^*\mathbb{R}^n)$ such that

$$A = Op_h^L(a) + O(h^\infty)_{\Psi^{-\infty}}, \qquad [Op_h^L(a)u](x) := \frac{1}{(2\pi h)^d} \int e^{\frac{i}{h}\langle x-y,\xi\rangle} a(x,\xi)u(y)\,dy\,d\xi.$$

Here, the integral in $Op_h^L(a)u$, can be understood as an iterated integral when $u \in \mathscr{S}(\mathbb{R}^n)$ and it is not hard to check that operators in $\Psi_\delta^m(\mathbb{R}^n)$ are bounded on $\mathscr{S}(\mathbb{R}^n)$ and $\mathscr{S}'(\mathbb{R}^n)$.

Remark 3.1.7 The L in the notation Op_h^L stands for the left quantization. There are many other standard choices of quantization including the Weyl quantization. We refer the reader to [Zwo12, Chap. 4] for more information.

As with symbols, we sometimes omit the space $\mathbb{R}^n$ from the notation and define $\Psi_\delta^\infty := \bigcup_m \Psi_\delta^m$, $\Psi_\delta^{-\infty} := \bigcap_m \Psi_\delta^m$. We also define $\Psi_\delta^{\mathrm{comp}}$ to be those $A \in \Psi_\delta^{-\infty}$ such that

$$A = Op_h^L(a) + O(h^\infty)_{\Psi^{-\infty}}$$

for some $a \in S_\delta^{\mathrm{comp}}$. Furthermore, we sometimes write $\Psi_0^m = \Psi^m$ i.e. omit the $\delta = 0$ from spaces of operators below. In what follows we will need the following result of [Zwo12, Theorems 4.14, 4.17] that explains the result of composition of two pseudodifferential operators.

Lemma 3.1.8 (Compositions) *Let* $0 \le \delta < \frac{1}{2}$ *and suppose that* $a \in S_\delta^{m_1}(T^*\mathbb{R}^n)$, $b \in S_\delta^{m_2}(T^*\mathbb{R}^n)$. *Then, there is* $c \in S_\delta^{m_1+m_2}(T^*\mathbb{R}^n)$ *such that*

$$Op_h^L(a)Op_h^L(b) = Op_h^L(c),$$

and for all N

$$\left(c(x,\xi) - \sum_{k=0}^{N-1} \frac{h^k}{k!}(i\langle D_\xi, D_y\rangle)^k (a(x,\xi)b(y,\eta))\Big|_{\substack{y=x\\\eta=\xi}}\right) \in h^{N(1-2\delta)} S_\delta^{m_1+m_2-N}. \qquad (3.1)$$

When (3.1) holds, we will write

$$c(x,\xi) \sim \sum_{k=0}^{\infty} \frac{h^k}{k!}(i\langle D_\xi, D_y\rangle)^k (a(x,\xi)b(y,\eta))\Big|_{\substack{y=x\\\eta=\xi}}.$$

We also recall the following asymptotic completeness lemma known as the Borel summation lemma (see e.g [Zwo12, Theorem 4.15]).

Lemma 3.1.9 (Asymptotic completeness) *Suppose that for $j = 0, 1, \ldots, c_j \in h^{-2\delta j} S_\delta^{m-j}$. Then there is $c \in S_\delta^m$ such that*

$$c \sim \sum_{j=0}^{\infty} h^j c_j.$$

3.1.2 Pseudodifferential Operators on a Compact Manifold

Next, let M be a compact, smooth manifold of dimension n. The phase space for a compact manifold is given by T^*M and we typically use coordinates $(x, \xi) \in T^*M$ where $x \in M$ and $\xi \in T_x^*M$.

Definition 3.1.10 (*Pseudodifferential operator on M*) We say that $A \in \Psi_\delta^m(M)$ if, for any local coordinate map $\psi : U \to \mathbb{R}^n$, and $\chi \in C_c^\infty(U)$,

$$\psi^* \chi A \chi [(\psi)^{-1}]^* \in \Psi_\delta^m(\mathbb{R}^n)$$

and for all $\chi_1, \chi_2 \in C_c^\infty(M)$ with $\operatorname{supp} \chi_1 \cap \operatorname{supp} \chi_2 = \emptyset$,

$$\chi_1 A \chi_2 = O(h^\infty)_{\Psi - \infty}.$$

Let g be a any smooth Riemannian metric on M and for $(x, \xi) \in T^*M$ let $\langle \xi \rangle := (1 + |\xi|_g^2)^{1/2}$. The symbol class $S_\delta^m(T^*M)$ is defined in the same way as $S_\delta^m(T^*\mathbb{R}^n)$ is defined in Definition 3.1.2 after replacing $\mathbb{R}^n$ with M.

Remark 3.1.11 Note that the spaces $S_\delta^m(T^*M)$ are independent of the choice of g and hence depend only on the smooth structure on M.

We also recall that there is a quantisation map $Op_h : S_\delta^m(T^*M) \to \Psi^\delta(M)$ such that for all $A \in \Psi_\delta^m(M)$, there is $a \in S_\delta^m(T^*M)$ such that

$$A = Op_h(a) + O(h^\infty)_{\Psi - \infty}.$$

For $s \in \mathbb{R}$, we define the semiclassical Sobolev space $H_h^s(M)$ as

$$H_h^s(M) := \{u \in \mathcal{D}'(M) : Op_h(\langle \xi \rangle^s) u \in L^2\}, \qquad \|u\|_{H_h^s(M)} := \|Op_h(\langle \xi \rangle^s) u\|_{L^2}.$$

Here, $\mathcal{D}'(M)$ denotes the space of distributions on M; i.e. the dual of $C^\infty(M)$.

Remark 3.1.12 Note that although the norm $\| \cdot \|_{H_h^s(M)}$ depends on the choice of the metric, g, the space does not and all norms are equivalent for any other choice of g.

3.1.3 Symbol Map

We now recall the most important, basic properties of the pseudodifferential calculus [DZ19, Appendix E].

Theorem 3.1.13 (Symbol map) *There is a map*

$$\sigma_{m,\delta} : \Psi_\delta^m(M) \to S_\delta^m(T^*M)/h^{1-2\delta}S_\delta^{m-1}(T^*M)$$

such that the following holds.

(1) Suppose that $A \in \Psi_\delta^m$ and $\sigma_{m,\delta}(A) = 0$. Then $A \in h^{1-2\delta}\Psi_\delta^{m-1}$.
(2) Suppose that $A \in \Psi_\delta^m$. Then, $A^* \in \Psi_\delta^m$ and $\sigma_{m,\delta}(A^*) = \overline{\sigma_{m,\delta}(A)}$.
(3) Let $A \in \Psi_\delta^{m_1}$ and $B \in \Psi_\delta^{m_2}$. Then $AB \in \Psi_\delta^{m_1+m_2}$ and

$$\sigma_{m_1+m_2,\delta}(AB) = \sigma_{m_1,\delta}(A)\sigma_{m_2,\delta}\sigma(B).$$

(4) Let $A \in \Psi_\delta^{m_1}$ and $B \in \Psi_\delta^{m_2}$. Then $[A, B] \in h^{1-2\delta}\Psi_\delta^{m_1+m_2-1}$ and

$$\sigma_{m_1+m_2-1,\delta}(h^{2\delta-1}[A, B]) = -ih^{2\delta}\{\sigma_{m_1,\delta}(A), \sigma_{m_2,\delta}(B)\},$$

where $\{a, b\}$ denotes the Poisson bracket of a and b.

Remark 3.1.14 Usually, we will write σ for the symbol map, leaving the m, δ implicit.

Finally, we record the boundedness properties of pseudodifferential operators [Zwo12, Proof of Thoerem 13.13].

Lemma 3.1.15 (Boundedness properties) *Let $A \in \Psi_\delta^m(M)$. Then for any $s \in \mathbb{R}$ there is $C > 0$ such that for $0 < h < 1$,*

$$\|A\|_{H_h^s(M)\to H_h^{s-m}(M)} \le C.$$

Furthermore, for $A \in \Psi_\delta^0(M)$, there is $C > 0$ such that

$$\|A\|_{L^2(M)\to L^2(M)} \le \sup_{T^*M} |\sigma(A)| + Ch^{1-2\delta}.$$

3.2   Wavefront Set and Microsupport

Before proceeding to properties of pseudodifferential operators such as ellipticity, we introduce the wavefront set of a pseudodifferential operator. Before doing so, it will be convenient to introduce the *fiber radially compactified*, $\overline{T}^*M$. This is a manfold with interior given by T^*M and

$$\partial\overline{T}^*M \cong S^*M.$$

We call S^*M, *fiber infinity*.

We now describe a neighborhood basis for each point $(x_0, \xi_0) \in \overline{T}^*M$. If $(x_0, \xi_0) \in T^*M$, then, the neighborhoods of (x_0, ξ_0) are the usual neighborhoods in T^*M. On the other hand, if $(x_0, \xi_0) \in S^*M = \partial\overline{T}^*M$, then a neighborhood basis is given as follows,

$$U_\varepsilon := \left\{ (x, \xi) \in \overline{T}^*M \ : \ |x - x_0| < \varepsilon, \ \left|\tfrac{\xi}{|\xi|} - \xi_0\right| < \varepsilon, \ |\xi| \geq \varepsilon^{-1} \right\}.$$

We can now define the essential support of a symbol and the wavefront set of a pseudodifferential operator. These notions codify the idea of the 'support' in phase space of a pseudodifferential operator.

Definition 3.2.1 (*Essential support*) Let $a \in S^m_\delta$. For $(x_0, \xi_0) \in \overline{T}^*M$, we say that $(x_0, \xi_0) \notin$ ess supp(a) if there is an h-independent neighborhood, U of (x_0, ξ_0) such that for all $\alpha, \beta \in \mathbb{N}^n$, and $N \in \mathbb{R}$, there is $C_{\alpha\beta N} > 0$ such that for $0 < h < 1$,

$$|\partial_x^\alpha \partial_\xi^\beta a(x, \xi)| \leq C_{\alpha\beta N} h^N \langle\xi\rangle^{-N}, \qquad (x, \xi) \in U \cap T^*M.$$

Definition 3.2.2 (*Wavefront set*) Let $A \in \Psi^m_\delta$. For $(x_0, \xi_0) \in \overline{T}^*M$, we say that $(x_0, \xi_0) \notin$ WF$_h(A)$ if there is $a \in S^m_\delta$ such that $(x_0, \xi_0) \notin$ ess supp(a) and

$$A = Op_h(a) + O(h^\infty)_{\Psi-\infty}.$$

It is easy to see from the definition that for any $A \in \Psi^m$, WF$_h(A) \subset \overline{T}^*M$ is closed.

The crucial feature of the wavefront set is contained in the following lemma [DZ19, (E.2.5)].

Lemma 3.2.3 *Suppose that* $A \in \Psi^{m_1}_\delta$, $B \in \Psi^{m_2}_\delta$. *Then,*

$$WF_h(AB) \subset WF_h(A) \cap WF_h(B).$$

Remark 3.2.4 Let $A \in \Psi^m$. Because of Lemma 3.2.3, one may think of WF$_h(A)$ as the set on which A 'lives'; i.e. Au contains no information about the parts of u which are not in

$\mathrm{WF_h}(A)$. We will see in the next section (see Lemma 3.3.2) that when $|\sigma(A)(x, \xi)| > c\langle \xi \rangle^m$ for $(x, \xi) \in U \subset T^*M$, then Au encodes all the information about the function u on U.

Definition 3.2.5 (*Wavefront set of a distribution*) For a distribution $u \in \mathcal{D}'(M)$ such that there are $N > 0$ and $C > 0$ with

$$\|u\|_{H_h^{-N}} \leq Ch^{-N},$$

we say that $\rho_0 \in \overline{T}^*M$ is not in the wavefront set of u, and write $\rho_0 \notin \mathrm{WF_h}(u)$, if there is $A \in \Psi^0(M)$ such that $\sigma(A)(\rho_0) = 1$ and for all $N > 0$ there is $C_N > 0$ such that

$$\|Au\|_{H_h^N} \leq C_N h^N.$$

Remark 3.2.6 We note that for the coherent state $u_j = h^{-\frac{n}{4}} e^{-\frac{|x - x_j|^2}{2h} + \frac{i}{h}\langle x - x_j, \xi_j \rangle}$, defined in (1.1), we have

$$\mathrm{WF_h}(u_j) = (x_j, \xi_j).$$

Furthermore, for any tempered function u, the pointwise localized functions $v_j = \chi(h^{-1}(x - x_j))u(x)$, as in (1.2), satisfy

$$\mathrm{WF_h}(\chi(h^{-1}(x - x_0))u) \subset T^*_{x_0}M$$

For the geodesic beam localizer, G_j, introduced in the introduction Chap.1, we have

$$\mathrm{WF_h}(G_j) \subset \bigcup_{|t| \leq T} \exp(t H_{|\xi|_g})(x_j, \xi_j).$$

for some $T >$. Finally, for the zonal harmonic type localizer Z_j, and any tempered function, u, we have

$$\mathrm{WF_h}(Zu) \subset \bigcup_{|t| \leq T} \exp(t H_{|\xi|_g})(T^*_x M).$$

The wavefront set, by definition, is an h-independent subset of $\overline{T}^*M$. In some cases, we will need a finer notion of 'support' that allows for h-dependent sets. This is the purpose of the microsupport.

Definition 3.2.7 (*Microsupport*) For a pseudodifferential operator $A \in \Psi_\delta^{\mathrm{comp}}(M)$, we say that A is microsupported in a family of sets $\{V(h)\}_h$ and write $\mathrm{MS_h}(A) \subset V(h)$ if

$$A = Op_h(a) + O(h^\infty)_{\Psi - \infty}$$

and for all α, N, there exists $C_{\alpha, N} > 0$ so that

$$\sup_{(x,\xi)\in T^*M\setminus V(h)} |\partial_{x,\xi}^\alpha a(x,\xi)| \le C_{\alpha,N} h^N.$$

For $B(h) \subset T^*M$, will also write $\mathrm{MS}_h(A) \cap B(h) = \emptyset$ for $\mathrm{MS}_h(A) \subset (B(h))^c$.

Note that the notation $\mathrm{MS}_h(A) \subset V(h)$ is a shortening for $\mathrm{MS}_h(A) \subset \{V(h)\}_h$.

Remark 3.2.8 (Microsupports of various localizers) We now describe the microsupports of the structured and unstructured localizers (see Figs. 1.2 and 1.1 respectively). We note that for the coherent state u_j defined in (1.1), we have

$$\mathrm{MS}_h(u_j) \subset B((x_j,\xi_j), h^\delta)$$

for any $\delta < \frac{1}{2}$. Furthermore, for any tempered function u, the pointwise localized functions $v_j = \chi(h^{-1}(x - x_j))u(x)$, as in (1.2), satisfy

$$\mathrm{MS}_h(\chi(h^{-1}(x - x_0))u) \subset B\big(T_{x_0}^*M, h^\delta\big)$$

for any $\delta < \frac{1}{2}$. For the geodesic beam localizer, G_j, introduced in (1.3), we have

$$\mathrm{MS}_h(G_j) \subset B\Big(\bigcup_{|t|\le T} \exp(t H_{|\xi|_g})(x_j,\xi_j), h^\delta\Big),$$

for any $\delta < \frac{1}{2}$ and some $T > 0$. Finally, for the zonal harmonic type localizer Z introduced in (1.4) and any tempered function, u, we have

$$\mathrm{MS}_h(Zu) \subset \Big(\bigcup_{|t|\le T} \exp(t H_{|\xi|_g})(T_x^*M), h^\delta\Big),$$

for any $\delta < \frac{1}{2}$. Indeed, the Z localizers provide microlocalization which is even finer than the microssuport indicates.

Since the notion of microsupport is less standard than the wavefront set, we include here proofs of various important properties.

Lemma 3.2.9 *Let* $0 \le \delta < \frac{1}{2}$ *and* $\delta' > \delta$, $c > 0$. *Suppose that* $A \in \Psi_\delta^{\mathrm{comp}}(M)$ *and that* $MS_h(A) \subset V(h)$. *Then,*

$$MS_h(A) \subset \Big\{(x,\xi) : d\big((x,\xi), V(h)^c\big) \le ch^{\delta'}\Big\}.$$

Proof Let $A = Op_h(a) + O(h^\infty)_{\psi-\infty}$. Suppose that

$$2r(h) := d\big(\rho_1, V(h)^c\big) \le ch^{\delta'}$$

and let $\rho_0 \in V(h)^c$ with $d(\rho_1, \rho_0) \leq r(h)$. Then, for any $N > 0$,

$$|\partial^\alpha a(\rho_1)| \leq \sum_{|\beta| \leq N-1} |\partial^{\alpha+\beta} a(\rho_0)| |r(h)|^{|\beta|} + C_{|\alpha|+N} \sup_{|k| \leq |\alpha|+N, T^*M} |\partial^k a| r(h)^N$$

$$\leq \sum_{|\beta| \leq N-1} \sup_{V^c} |\partial^{\alpha+\beta} a(\rho)| |r(h)|^{|\beta|} + C_{\alpha N} h^{-N\delta} r(h)^N$$

$$\leq C_{\alpha N M} h^M + C_{\alpha N} h^{-N\delta} r(h)^N$$

So, letting $N \geq M(\delta' - \delta)^{-1}$,

$$|\partial^\alpha a(\rho_1)| \leq C_{\alpha M} h^M.$$

$\square$

Lemma 3.2.10 *Let* $0 \leq \delta < \frac{1}{2}$ *and* $A, B \in \Psi_\delta^{\mathrm{comp}}(M)$. *Suppose that* $MS_h(A) \subset V(h)$ *and* $MS_h(B) \subset W(h)$.

(1) The statement $MS_h(A) \subset V(h)$ *is well defined. In particular, it does not depend on the choice of quantization procedure.*
(2) $MS_h(AB) \subset V(h) \cap W(h)$
(3) $MS_h(A^*) \subset V(h)$
(4) If $V(h) = \emptyset$, *then* $WF_h(A) = \emptyset$.
(5) If $A = Op_h(a) + O(h^\infty)_{\Psi-\infty}$, *then* $MS_h(a) \subset \mathrm{supp}\, a$.

Proof The proofs of 1–3 are nearly identical, relying on the asymptotic expansion for, respectively, the change of quantization, composition, and adjoint so we write the proof for only (2). Write

$$A = Op_h(a) + O(h^\infty)_{\Psi-\infty}, \qquad B = Op_h(b) + O(h^\infty)_{\Psi-\infty}.$$

Then,

$$Op_h(a) Op_h(b) = Op_h(a\#b) + O(h^\infty)_{\Psi-\infty}$$

where

$$a\#b(x, \xi) \sim \sum_j h^j L_{2j} a(x, \xi) b(y, \eta) \Big|_{\substack{x=y \\ \xi=\eta}}$$

and L_{2j} are differential operators of order $2j$. Suppose that $MS_h(A) \subset V$. Then, for any $N > 0$.

$$\sup_{V^c} |\partial^\alpha a| \leq C_{\alpha N} h^N.$$

So, choosing $M > (N + \delta|\alpha|)(1 - 2\delta)^{-1}$,

$$\left| \partial^\alpha a \# b \right| \leq \left| \partial^\alpha \sum_{j < M} h^j L_{2j} a(x, \xi) b(y, \eta) \Big|_{\substack{x=y \\ \xi=\eta}} \right| + C_{\alpha M} h^{M(1-2\delta) - |\alpha|\delta} \leq C_{\alpha N} h^N$$

In particular,

$$\sup_{V^c} \left| \partial^\alpha a \# b \right| \leq C_{\alpha N} h^N.$$

An identical argument shows

$$\sup_{W^c} \left| \partial^\alpha a \# b \right| \leq C_{\alpha N} h^N.$$

(4) follows from the definition since if $V(h) = \emptyset$, $a \in h^\infty S_\delta$.

(5) follows easily from the definition. $\qquad\qquad\qquad\qquad\qquad\qquad\qquad\square$

Definition 3.2.11 (*Microsupport of a function*) For a distribution $u \in \mathcal{D}'(M)$ such that there are $N > 0$ and $C > 0$ with

$$\|u\|_{H_h^{-N}} \leq C h^{-N},$$

we say that u is microsupported in $V(h)$, and write $\mathrm{MS}_h(u) \subset V(h)$, if there is $A \in \Psi_\delta^{\mathrm{comp}}(M)$ with $\mathrm{MS}_h(A) \subset V(h)$ and for any $N > 0$ there is $C_N > 0$ such that

$$\|(I - A)u\|_{H_h^N} \leq C_N h^N.$$

3.3   Ellipticity and Inverses

We now define the notion of ellipticity for pseudodifferential operators.

Definition 3.3.1 (*Ellipticity*) Let $A \in \Psi^m(M)$. For $(x_0, \xi_0) \in \overline{T}^*M$, we say that A *is elliptic at* (x_0, ξ_0), and write $(x_0, \xi_0) \in \mathrm{Ell}(A)$, if there is a neighborhood, $U \subset \overline{T}^*M$ of (x_0, ξ_0) and $c > 0$ such that

$$|\sigma(A)(x, \xi)| \geq c \langle \xi \rangle^m, \qquad (x, \xi) \in U \cap T^*M$$

It is easy to see from the definition that for any $A \in \Psi^m(M)$, $\mathrm{Ell}(A) \subset \overline{T}^*M$ is open.

Ellipticity gives an appropriate conditions which guarantee that A is invertible on a subset of $\overline{T}^*M$ in the following sense.

Lemma 3.3.2 (Elliptic parametrix) *Suppose that* $A \in \Psi_\delta^{m_1}$ *and* $B \in \Psi_\delta^{m_2}$ *with* $\mathrm{WF}_h(B) \subset \mathrm{Ell}(A)$. *Then there are* $E_L, E_R \in \Psi_\delta^{m_2 - m_1}$ *such that*

$$B = E_L A + O(h^\infty)_{\Psi^{-\infty}}, \qquad B = A E_R + O(h^\infty)_{\Psi^{-\infty}}.$$

As with many constructions in semiclassical analysis, this lemma is proved by an iterative construction. The nonlinear part of the construction is done by solving a top order equation, and then each successive iteration involves only the solution of a linear equation. In the case of the elliptic parametrix construction, this is particularly simple since the equations involved are algebraic.

Proof Let $e = \sigma(B)/\sigma(A)$. Then, since $\mathrm{WF_h}(B) \subset \mathrm{Ell}(A)$, $|\sigma(A)| > c > 0$ on $\mathrm{supp}\,\sigma(B)$, and hence, $e \in S_\delta^{m_2 - m_1}$. Putting $E_{L,0} := Op_h(e)$, we have

$$\sigma_{m_2,\delta}(E_{L,0}A - B) = 0,$$

and therefore,

$$E_{L,0}A = B + h^{1-2\delta}R_1,$$

with $R_1 \in \Psi_\delta^{m_1 - 1}$.

Suppose we have found e_j, $i = 0, 1, \ldots, N-1$, $e_j \in S_\delta^{m_2 - m_1 - j}$ such that $\mathrm{supp}\,e_j \subset \mathrm{WF_h}(B)$, and, with $E_{L,N-1} := \sum_{j=0}^{N-1} h^{j(1-2\delta)} Op_h(e_j)$, we have

$$E_{L,N-1}A = B + h^{N(1-2\delta)}R_N, \tag{3.2}$$

for some $R_N \in \Psi_\delta^{m_2 - N}$. Now, since $\mathrm{supp}\,e_i \subset \mathrm{WF_h}(B)$, $\mathrm{WF_h}(E_{L,N-1}) \subset \mathrm{WF_h}(B)$ and hence,

$$\mathrm{WF_h}(R_N) = \mathrm{WF_h}(h^{-N(1-2\delta)}(B - E_{L,N-1}A)) \subset \mathrm{WF_h}(B).$$

Therefore $\mathrm{WF_h}(R_N) \subset \mathrm{Ell}(A)$ and hence $e_N := -\sigma(R_N)/\sigma(A) \in S_\delta^{m_2 - N - m_1}$ and

$$(E_{L,N-1} + h^{N(1-2\delta)} Op_h(e_N))A - B = h^{N(1-2\delta)}(R_N + Op_h(e_N)A) \in h^{N(1-2\delta)}\Psi^{m_2 - N},$$

and

$$\sigma_{m_2 - N,\delta}(R_N + Op_h(e_N)A) = 0.$$

Therefore,

$$(E_{L,N-1} + h^{N(1-2\delta)} Op_h(e_N))A - B = h^{(N+1)(1-2\delta)}R_{N+1},$$

for some $R_{N+1} \in \Psi^{m_2 - N - 1}$. In particular, putting $E_{L,N} = \sum_{j=0}^{N} h^j Op_h(e_j)$, we have (3.2) with $N - 1$ replaced by N. In particular, there are $e_j \in \Psi_\delta^{m_2 - m_2 - j}$ for $j = 0, 1, \ldots$ such that (3.2) holds for any N. Setting $E_L \sim \sum_j h^{j(1-2\delta)} Op_h(e_j)$, completes the proof of the first equality.

The proof of the second equality is nearly identical and we leave the details to the reader. $\qquad\square$

We will need a finer version of this lemma in some of our applications. We record and prove it here.

Lemma 3.3.3 *Let* $0 \leq \delta < \frac{1}{2}$, $P \in \Psi^m(M)$ *with principal symbol* p, *and*

$$\chi \in S_\delta^{\text{comp}} \cap C_c^\infty(T^*M; [-C_0 h^{1-2\delta}, 1 + C_0 h^{1-2\delta}])$$

be so that there exist $c, h_1 > 0$ *with*

$$\text{supp}\,\chi \subset \{|p| \geq ch^\delta \,,\, |p| + |dp| > c\}$$

for $0 < h \leq h_1$. *Then for all* $\tilde{\chi} \in S_\delta \cap C_c^\infty(T^*M; [0, 1])$ *with* $\tilde{\chi} \equiv 1$ *on* $\text{supp}\,\chi$, *there exists* $b \in h^{-\delta}\S_\delta$ *such that*

$$Op_h(\chi) = Op_h(b)\,Op_h(\tilde{\chi})\,P + O(h^\infty)_{\Psi-\infty}$$

Moreover, the S_δ *seminorms of* b *are uniform for* $\tilde{\chi}$, χ *in bounded subsets of* S_δ, *and for* Θ *in bounded subsets of* C^∞.

Proof First, let $\psi \in C_c^\infty(\mathbb{R})$ with $\psi \equiv 1$ on $[-1, 1]$. Then, using the standard elliptic parametrix construction (see the proof of Lemma 3.3.2) there exists $b_1 \in S_\delta^{\text{comp}}$ with $\sup |b_1| \leq 2c^{-1} + C_1 h^{1-2\delta}$ such that

$$Op_h(\chi)\,Op_h\left(1 - \psi\left(\tfrac{2}{c}p\right)\right) = Op_h(b_1)\,Op_h(\tilde{\chi})\,P + O(h^\infty)_{\Psi-\infty}. \tag{3.3}$$

Next, we show that there exists $b_2 \in S_\delta^{\text{comp}}$ with $\sup |b_2| \leq c^{-1}h^{-\delta} + C_1 h^{1-3\delta}$ so that

$$Op_h(\chi)\,Op_h\left(\psi\left(\tfrac{2}{c}p\right)\right) = Op_h(b_2)\,Op_h(\tilde{\chi})\,P + O(h^\infty)_{\Psi-\infty}. \tag{3.4}$$

Using that $|p| \geq ch^\delta$ on $\text{supp}\,\chi$ one can carry out an elliptic parametrix construction in the second microlocal calculus associated to $p = 0$. Using a partition of unity, since $|dp| > \frac{c}{2}$ on $\text{supp}\,\chi \cap \text{supp}\,\psi\left(\tfrac{2}{c}p\right)$ we may assume that there exist an h-independent neighborhood V_0 of $\text{supp}\,\chi$, $V_1 \subset T^*\mathbb{R}^n$ a neighborhood of 0, and a symplectomorphism $\kappa : V_1 \to V_0$ so that $\kappa^* p = \xi_1$. Let U be a microlocally unitary FIO quantizing κ (see [Zwo12, Theorem 11.6] for the construction and properties). Then

$$\mathbf{P} := U^* P U = hD_{x_1} + hOp_h^L(\mathbf{r}),$$

with $\mathbf{r} \in S^{\text{comp}}(\mathbb{R}^n)$ and Op_h^L denotes the *left* quantization of $\mathbf{r}$. Moreover, there exist $\mathbf{a}, \tilde{\mathbf{a}} \in S_\delta^{\text{comp}}(T^*\mathbb{R}^n)$ so that

$$Op_h^L(\mathbf{a}) = U^* Op_h(\chi)\,Op_h\left(\psi\left(\tfrac{2}{c}p\right)\right)U$$

and

$$Op_h^L(\tilde{\mathbf{a}}) = U^* Op_h(\tilde{\chi})U$$

with $\text{supp}\,\mathbf{a} \subset \{|\xi_1| \geq ch^\delta\}$ and $\tilde{\mathbf{a}} \equiv 1$ on $\text{supp}\,\mathbf{a}$. Now, for $\mathbf{b} \in S_\delta^{\text{comp}}(T^*\mathbb{R}^n)$ supported on $|\xi_1| \geq ch^\delta$,

$$|\partial_x^\alpha \partial_\xi^\beta (\xi_1^{-1} \mathbf{b})| \leq C_{\alpha\beta} h^{-(|\beta|+|\alpha|)\delta} |\xi_1|^{-1}.$$

Let $\mathbf{b}_0 = \mathbf{a}/\xi_1$. Then $\mathbf{b}_0 \in h^{-\delta} S_\delta^{\mathrm{comp}}$ and

$$\sup |\mathbf{b}_0| \leq c^{-1} h^{-\delta}.$$

Observe that by Lemma 3.1.8

$$Op_h^L(\mathbf{b}_0) Op_h^L(\tilde{\mathbf{a}})\mathbf{P} = Op_h^L(\mathbf{a}) + Op_h^L(\mathbf{e}_1) + O(h^\infty)_{\Psi-\infty}$$

with $\mathrm{supp}\,\mathbf{e}_1 \subset \{|\xi_1| \geq ch^\delta\}$ and, since $\tilde{\mathbf{a}} \equiv 1$ on $\mathrm{supp}\,\mathbf{b}_0$,

$$\mathbf{e}_1 \sim \sum_{|\alpha|\geq 1} \frac{h^{|\alpha|} i^{|\alpha|}}{\alpha!} D_x^\alpha(\mathbf{b}_0) D_\xi^\alpha(\xi_1) + \sum_{|\alpha|\geq 0} \frac{h^{|\alpha|+1} i^{|\alpha|}}{k!} D_x^\alpha(\mathbf{b}_0) D_\xi^\alpha(\mathbf{r}).$$

In particular, $\mathbf{e}_1 \in h^{1-2\delta} S_\delta^{\mathrm{comp}}$. Then, setting $\mathbf{b}_\ell = -\mathbf{e}_\ell/\xi_1 \in h^{\ell(1-2\delta)-\delta} S_\delta^{\mathrm{comp}}$, and

$$Op_h^L(\mathbf{e}_{\ell+1}) := Op_h^L(\mathbf{b}_\ell) Op_h^L(\tilde{\mathbf{a}})\mathbf{P} + Op_h^L(\mathbf{e}_\ell) + O(h^\infty)_{\Psi-\infty}$$

we have $\mathbf{e}_{\ell+1} \in h^{(\ell+1)(1-2\delta)} S_\delta^{\mathrm{comp}}$ with $\mathrm{supp}\,\mathbf{e}_{\ell+1} \subset \{|\xi_1| \geq ch^\delta\}$. In particular, putting $\mathbf{b} \sim \sum_\ell \mathbf{b}_\ell$,

$$Op_h^L(\mathbf{b}) Op_h^L(\tilde{\mathbf{a}})\mathbf{P} = Op_h^L(\mathbf{a}) + O(h^\infty)_{\Psi-\infty}.$$

It follows that

$$\begin{aligned}
U Op_h^L(\mathbf{b}) U^* Op_h(\tilde{\chi}) P &= U Op_h^L(\mathbf{b}) U^* U Op_h^L(\tilde{\mathbf{a}}) U^* UPU^* + O(h^\infty)_{\Psi-\infty}\\
&= U Op_h^L(\mathbf{b}) Op_h^L(\tilde{\mathbf{a}}) PU^* + O(h^\infty)_{\Psi-\infty}\\
&= U Op_h^L(\mathbf{a}) U^* + O(h^\infty)_{\Psi-\infty}\\
&= Op_h(\chi) Op_h(\psi(\tfrac{2}{c}p)) + O(h^\infty)_{\Psi-\infty}.
\end{aligned}$$

In particular, there exists $b_2 \in h^{-\delta} S_\delta^{\mathrm{comp}}(T^*M)$ with $\sup |b_2| \leq c^{-1} h^{-\delta} + C_1 h^{1-3\delta}$ so that

$$Op_h(b_2) = U Op_h^L(\mathbf{b}) U^* + O(h^\infty)_{\Psi-\infty}.$$

Therefore, as claimed in (3.4) that

$$Op_h(\chi) Op_h(\psi(\tfrac{2}{c}p)) = Op_h(b_2) Op_h(\tilde{\chi}) P + O(h^\infty)_{\Psi-\infty},$$

for all χ supported in V_0 and some suitable b_2 with $\|Op_h(b_2)\| \leq 2c^{-1} h^{-\delta}$. Writing $b = b_1 + b_2$, we obtain using (3.3) and (3.4) that

$$Op_h(\chi) = Op_h(b) Op_h(\tilde{\chi}) P + O(h^\infty)_{\Psi-\infty}$$

as claimed. $\qquad\square$

3.4   Egorov's Theorem

Finally, we record a version of Egorov's theorem to the Ehrefest time which will be used when passing from local analytical information to global dynamical information. For $t \in \mathbb{R}$, let $\varphi_t : T^*M \setminus \{0\} \to T^*M \setminus \{0\}$ be given by

$$\varphi_t := \exp(t \, H_{|\xi|_g}).$$

Then φ_t is the geodesic flow. Define the maximal expansion rate

$$\Lambda_{\max} := \sup_{\rho \in S^*M} \limsup_{t \to \infty} \frac{1}{t} \log \|d\varphi_t(\rho)\|.$$

Then, $0 \leq \Lambda_{\max} < \infty$ and we define the Ehrenfest time

$$T_e(h) := \frac{\log h^{-1}}{2\Lambda_{\max}}.$$

Here, if $\Lambda_{\max} = 0$, we take $T_e(h) := M \log h^{-1}$ for an arbitrary $M > 0$. We now state a consequence of [DG14, Proposition 3.9].

Theorem 3.4.1 (Egorov's Theorem to the Ehrenfest time) *Let $0 \leq \delta < \frac{1}{2}$ and $\varepsilon > 0$. Then there is $\varepsilon_1 > 0$ such that for all $A \in \Psi_\delta^{\mathrm{comp}}(M)$ with $WF_h(A) \subset \{1 - \varepsilon_1 < |\xi|_g < 1 + \varepsilon_1\}$, putting*

$$A(t) := e^{-ith\Delta_g} A e^{ith\Delta_g}.$$

we have for $0 < \gamma \leq \frac{1}{2} - \delta$ all $0 \leq t \leq (2\gamma - \varepsilon) \log h^{-1}$, $A(t) \in \Psi_{\delta + \gamma}^{\mathrm{comp}}$,

$$\sigma(A(t)) = \sigma(A) \circ \varphi_t,$$

and

$$MS_h(A(t)) \subset \varphi_{-t}(MS_h(A)).$$

Remark 3.4.2 The Egorov theorem in [DG14, Proposition 3.9] is stated on a potentially non-compact manifold for $A \in \Psi_0$. However, it is easy to see from the proof of this proposition that one can apply it to $A \in \Psi_\delta$ by reducing the time of propagation accordingly.

Basic Properties of Eigenfunctions and Eigenvalues

We are interested in studying the behavior of solutions to

$$(-\Delta_g - \lambda^2)u = f,$$

as $\lambda \to \infty$. It will be convenient to switch to the semiclassical Laplacian, putting $h = \frac{1}{\lambda}$ and multiplying by h^2 to obtain

$$(-h^2 \Delta_g - 1)u = h^2 f. \tag{4.1}$$

From now on, we study solutions to (4.1) as $h \to 0^+$.

In this chapter, we introduce the semiclassical Laplacian, $h^2 \Delta_g \in \Psi_h^2(M)$, with principal symbol $-|\xi|^2_{g(x)}$. We show that its eigenfunctions are smooth, its spectrum is discrete, and that one can build an orthonormal basis of $L^2(M)$ consisting of Laplace eigenfunctions. The proofs in this section are inspired by the presentation in [Zwo12, Sect. 14.3].

4.1 The Semiclassical Laplacian

The semiclassical Laplacian is the operator $h^2 \Delta_g$ where Δ_g is the Laplace operator defined in (2.1). We may define $h^2 \Delta_g : L^2(M) \to \mathcal{D}'(M)$ by $h^2 \Delta_g u : C^\infty(M) \to \mathbb{C}$ for $u \in L^2(M)$ as

$$h^2 \Delta_g u\,(v) := \int_M u \overline{(h^2 \Delta_g v)}.$$

Then, $h^2 \Delta_g : L^2(M) \to L^2(M)$ is an unbounded operator with domain $\mathrm{Dom}(h^2 \Delta_g) = C^\infty(M)$.

The operator $-h^2 \Delta_g$ is *positive definite* and *symmetric*, since

© The Author(s), under exclusive license to Springer Nature Switzerland AG 2023
Y. Canzani and J. Galkowski, *Geodesic Beams in Eigenfunction Analysis*, Synthesis
Lectures on Mathematics & Statistics, https://doi.org/10.1007/978-3-031-31586-2_4

$$\langle u, -h^2\Delta_g v\rangle_{L^2} = \int_M u\,\overline{-h^2\Delta_g v}\,\,dv_g = \int_M \langle h\nabla_g u, h\nabla_g v\rangle dv_g$$

$$= \int_M -h^2\Delta_g u\,\overline{v}\,\,dv_g = \langle -h^2\Delta_g u, v\rangle_{L^2}$$

for $u, v \in C^\infty(M)$. However, $h^2\Delta_g : L^2(M) \to L^2(M)$ is not self-adjoint since

$$\mathrm{Dom}(h^2\Delta_g)^* = H_h^2(M) \supsetneq C^\infty(M). \tag{4.2}$$

We will show in Proposition 4.3.1 that $-h^2\Delta_g$ admits a self-adjoint extension with $\mathrm{Dom}(h^2\Delta_g) = H_h^2(M)$. Note that $h^2\Delta_g : H_h^2(M) \to L^2(M)$ is bounded.

According to (2.2), in local coordinates $(x_1, \ldots, x_n)$, the Laplacian takes the form

$$-h^2\Delta_g = \sum_{i,j=1}^n g^{ij} hD_{x_i} hD_{x_j} + \frac{hi}{\sqrt{|\det g|}} \sum_{i,j=1}^n \partial_{x_i}\left(g^{ij}\sqrt{|\det g|}\right) hD_{x_j}.$$

It follows that $-h^2\Delta_g \in \Psi_h^2(M)$ and has principal symbol

$$\sigma(-h^2\Delta_g)(x, \xi) = \sum_{i,j=1}^n g^{ij}(x)\xi_i\xi_j = |\xi|^2_{g(x)}. \tag{4.3}$$

4.2   Smoothness and Existence of the Resolvent

Here, we prove that Laplace eigenfunctions are smooth and that the inverse of $-h^2\Delta_g - z :$ $H_h^2(M) \to L^2(M)$ exists for $z \in \mathbb{C}\backslash\mathbb{R}$. The existence of the resolvent will be the key tool in the proof of the Spectral Theorem, Proposition 4.3.2.

We first present a key lemma that will be used to prove both the smoothness of eigenfunctions and the existence of the resolvent.

Lemma 4.2.1 *If $|\Im z| > 0$, then for each $N \in \mathbb{N}$ there are $Q_N \in \Psi_h^{-2}(M)$ and $R_{N+1} \in \Psi_h^{-(N+1)}(M)$ such that*

$$Q_N(-h^2\Delta_g - z) = I - R_{N+1}. \tag{4.4}$$

Proof Let $\chi \in C^\infty(\mathbb{R}; [0, 1])$ with $\chi \equiv 1$ on $[-T, T]$ with $T > 0$ so that $T^2 > |z|$, and set

$$q_0(x, \xi) = \frac{1 - \chi(|\xi|)}{|\xi|^2 - z},$$

Note that $q_0 \in S^{-2}(T^*M)$ and set $Q_0 = Op_h(q_0) \in \Psi_h^{-2}(M)$. Then, for some $B_1 \in \Psi_h^{-1}(M)$,

$$Q_0 Op_h(|\xi|^2 - z) = I - Op_h(\chi(|\xi|)) + hB_1.$$

It follows that

$$Q_0(-h^2\Delta_g - z) = I + hB_1 + Q_0 hL_1 + Op_h(\chi(|\xi|)) =: I - R_1.$$

Finally, one sets

$$Q_N := \sum_{j=1}^{N} R_1^j Q_0 \in \Psi_h^{-2}(M)$$

and so

$$Q_N(-h^2\Delta_g - z) = \sum_{j=1}^{N} R_1^j Q_0(-h^2\Delta_g - z) = \sum_{j=1}^{N} R_1^j(I - R_1) = I - R_1^{N+1}.$$

Letting $R_{N+1} := R_1^{N+1} \in \Psi_h^{-(N+1)}(M)$ finishes the proof. $\qquad\qquad\square$

We are now ready to establish the smoothness of the Laplace eigenfunctions.

Proposition 4.2.2 (Smooth eigenfunctions) *Let $z \in \mathbb{C}$ and $u \in L^2(M)$ such that $(-h^2\Delta_g - z)u = 0$. Then, $u \in C^\infty(M)$. In particular, eigenfunctions of $-h^2\Delta_g$ are $C^\infty(M)$.*

Proof Let $u \in L^2(M)$ such that $(-h^2\Delta_g - z)u = 0$. By Lemma 4.2.1, for all $N > 0$ there exist $Q_N \in \Psi_h^{-2}(M)$ and $R_{N+1} \in \Psi_h^{-(N+1)}(M)$ such that

$$Q_N(-h^2\Delta_g - z) = I - R_{N+1}.$$

In particular, $(I - R_{N+1})u = 0$ and so $u = R_{N+1}u$. Since $R_{N+1} \in \Psi_h^{-(N+1)}(M)$, we know $R_{N+1} : L^2(M) \to H_h^{N+1}(M)$. Thus, $u \in H_h^{N+1}(M)$ for all N. By the Sobolev embedding we conclude $u \in C^\infty(M)$. $\qquad\qquad\square$

Finally, we prove the existence of the inverse of $-h^2\Delta_g - z$ for $z \in \mathbb{C}\backslash\mathbb{R}$.

Proposition 4.2.3 (Existence of the resolvent) *If $z \in \mathbb{C}\backslash\mathbb{R}$, then $-h^2\Delta_g - z : H_h^2(M) \to L^2(M)$ is invertible and*

$$\|(-h^2\Delta_g - z)^{-1}\|_{L^2 \to H_h^2} = O\left(1 + \frac{1 + |z|}{|\Im z|}\right).$$

Proof First, we claim that

$$\mathrm{Ker}(-h^2\Delta_g - z) = \{0\}. \tag{4.5}$$

To see this, let $u \in L^2(M)$ such that $(-h^2\Delta_g - z)u = 0$. Then, by Proposition 4.2.2 we have $u \in C^\infty(M)$. It follows that

$$0 = \langle (-h^2 \Delta_g - z)u, u \rangle_{L^2} = \langle u, (-h^2 \Delta_g - \bar{z})u \rangle_{L^2} = \langle u, (z - \bar{z})u \rangle_{L^2},$$

where we used that $u \in C^\infty(M)$, $-h^2 \Delta_g$ is symmetric, and $(-h^2 \Delta_g - z)u = 0$.

We conclude that $0 = 2i \Im z \|u\|_{L^2}^2$. Since $u \in C^\infty(M)$ and $\Im z \neq 0$, it follows that $u = 0$ as claimed in (4.5).

Next, we prove that $-h^2 \Delta_g - z : H_h^2(M) \to L^2(M)$ is surjective. To see this, let $u \in [(-h^2 \Delta_g - z)(C^\infty(M))]^\perp \subset L^2(M)$. Then, for all $v \in C^\infty(M)$,

$$0 = \langle u, (-h^2 \Delta_g - z)v \rangle_{L^2} = \langle (-h^2 \Delta_g - \bar{z})u, v \rangle_{L^2},$$

and so $(-h^2 \Delta_g - \bar{z})u = 0$ in $\mathcal{D}'(M)$. By Proposition 4.2.2 we know $u \in C^\infty(M)$ and so (4.5) yields that $u = 0$. In particular,

$$L^2(M) = \overline{(-h^2 \Delta_g - z)(C^\infty(M))} = (-h^2 \Delta_g - z)(H_h^2(M)), \tag{4.6}$$

where we used that $-h^2 \Delta_g - z : H_h^2(M) \to L^2(M)$ is a bounded operator and hence its image is closed.

Having established that $-h^2 \Delta_g - z : H_h^2(M) \to L^2(M)$ is surjective, we proceed to prove that there is $C > 0$ such that

$$\|(-h^2 \Delta_g - z)^{-1}v\|_{H_h^2} \leq \frac{C}{|\Im z|} \|v\|_{L^2}. \tag{4.7}$$

To prove (4.7) we start by noting that for all $u \in C^\infty(M)$

$$\Im \langle (-h^2 \Delta_g - z)u, u \rangle = -\Im z \|u\|_{L^2}^2.$$

Thus, for all $u \in C^\infty(M)$,

$$|\Im z| \|u\|_{L^2} \leq \|(-h^2 \Delta_g - z)u\|_{L^2}$$

It follows that for all $v \in (-h^2 \Delta_g - z)(C^\infty(M))$

$$\|v\|_{L^2} \geq |\Im z| \|((-h^2 \Delta_g - z)^{-1}v\|_{L^2}.$$

We claim that this implies that for all $v \in L^2(M)$

$$\|(-h^2 \Delta_g - z)^{-1}v\|_{L^2} \leq \frac{1}{|\Im z|} \|v\|_{L^2}. \tag{4.8}$$

To see (4.8), let $v \in L^2(M)$. Then, $v = (-h^2 \Delta_g - z)u$ for some $u \in H_h^2(M)$. Then, let $\{u_j\} \subset C^\infty(M)$ with $\lim_{j \to \infty} u_j = u$ in $H_h^2(M)$. It follows that $\lim_{j \to \infty}(-h^2 \Delta_g - z)u_j = (-h^2 \Delta_g - z)u$ since $-h^2 \Delta_g - z : H_h^2(M) \to L^2(M)$ is bounded. Therefore,

$$\|v\|_{L^2} = \|(-h^2 \Delta_g - z)u\|_{L^2} = \lim_{j \to \infty} \|(-h^2 \Delta_g - z)u_j\|_{L^2} \geq |\Im z| \lim_{j \to \infty} \|u_j\|_{L^2} = |\Im z| \|u\|_{L^2},$$

proving (4.8).

Finally, we proceed to strengthen (4.8) to prove (4.7). Notice that it suffices to show that for all $u \in H_h^2(M)$,

$$\|u\|_{H_h^2} \le C\|-h^2\Delta_g u\|_{L^2} + C\|u\|_{L^2}. \tag{4.9}$$

Indeed, to then derive (4.7) for $v \in L^2(M)$ one could set $u = (-h^2\Delta_g - z)^{-1}v$ and obtain from (4.8) that that

$$\|(-h^2\Delta_g - z)^{-1}v\|_{H_h^2} \le C\|v\|_{L^2} + C(|z| + \frac{1}{|\Im z|})\|v\|_{L^2},$$

as desired in (4.9).

To prove (4.9), note that by Lemma 4.2.1, there are $Q_1 \in \Psi_h^{-2}(M)$, $R_2 \in \Psi_h^{-2}(M)$ such that $Q_1(-h^2\Delta_g) = I - R_2$. Therefore, there is $C > 0$ such that

$$\begin{aligned}
\|u\|_{H_h^2} &= \|Op_h(\langle\xi\rangle^2)u\|_{L^2} \\
&\le \|Op_h(\langle\xi\rangle^2)Q_1(-h^2\Delta_g)u\|_{L^2} + \|Op_h(\langle\xi\rangle^2)R_2 u\|_{L^2} \\
&\le C\|(-h^2\Delta_g)u\|_{L^2} + C\|u\|_{L^2}. \tag{4.10}
\end{aligned}$$

Here, to obtain (4.10), we used that $Op_h(\langle\xi\rangle^2)Q_1 \in \Psi_h^0(M)$ and $Op_h(\langle\xi\rangle^2)R_2 \in \Psi_h^0(M)$ are bounded. This yields the bound in (4.9). $\qquad\square$

4.3 Spectral Theorem

The last section of this chapter is dedicated to proving that there is a self-adjoint extension for the Laplacian and that one can construct an orthonormal basis of $L^2(M)$ consisting of its eigenfunctions.

Proposition 4.3.1 (Self-adjoint extension) *The Laplacian $-h^2\Delta_g$ with domain $C^\infty(M)$ is essentially self-adjoint and its closure yields a self-adjoint extension of $-h^2\Delta_g$ with domain $H_h^2(M)$.*

Proof To prove the existence of the self-adjoint extension of $-h^2\Delta_g$, one uses the fact that, by (4.6),

$$\overline{(-h^2\Delta_g \pm i)(C^\infty(M))} = L^2(M).$$

Since $-h^2\Delta_g : L^2(M) \to L^2(M)$ is symmetric, this implies that $-h^2\Delta_g$ is essentially self-adjoint. In particular, its closure defines a self-adjoint extension of $-h^2\Delta_g$ whose domain is $H_h^2(M) = \text{Dom}((-h\Delta_g)^*)$ (see (4.2)). $\qquad\square$

We are finally ready to prove the Spectral Theorem for the Laplacian.

Proposition 4.3.2 (Spectral Theorem) *There exists a self-adjoint extension of* $-h^2\Delta_g$ *with domain* $H_h^2(M)$. *Furthermore, for each* h *there exists a smooth, orthonormal basis* $\{u_k(h)\}_k$ *of* $L^2(M)$ *and* $\{E_k(h)\}_k \subset [0, \infty)$ *with* $\lim_k E_k(h) = +\infty$ *such that*

$$-h^2\Delta_g u_k(s) = E_k(h)u_k(h).$$

Proof First, note that while $-h^2\Delta_g - z : H_h^2(M) \to L^2(M)$ is not self-adjoint, it is still normal. That is,

$$(-h^2\Delta_g - z)^*(-h^2\Delta_g - z) = (-h^2\Delta_g - z)(-h^2\Delta_g - z)^*.$$

Thus, $(-h^2\Delta_g - z)^{-1}$ is a normal operator as well.

Next, note that by Lemma 4.2.1 there are $Q_1 \in \Psi_h^{-2}(M)$, $R_2 \in \Psi_h^{-2}(M)$ such that $Q_1(-h^2\Delta_g - z) = I - R_2$ and so

$$(-h^2\Delta_g - z)^{-1} = Q_1 + R_2(-h^2\Delta_g - z)^{-1}.$$

In particular, by Proposition 4.2.2, the resolvent $(-h^2\Delta_g - z)^{-1}$ is a compact operator.

By the Spectral Function Theorem for compact operators [Zwo12, Theorem C.10], there exists an orthonormal basis $\{u_k(h)\}_k$ of $L^2(M)$ and $\{\mu_k(h)\}_k \subset [0, \infty)$ with $\lim_k \mu_k(h) = 0$ so that

$$(-h^2\Delta_g - i)^{-1}u_k(s) = \mu_k(h)u_k(h).$$

The result follows after setting $E_k(h) := \frac{1}{\mu_k(h)} + i$. The fact that $\{u_k(h)\}_k \subset C^\infty(M)$ follows from Proposition 4.2.2. $\qquad\square$

The Koch–Tataru–Zworski Approach to L^∞ Estimates

The idea behind the Koch–Tataru–Zworski [KTZ07] approach to L^p estimates is that every pseudodifferential operator P such that $p = \sigma(P)$ is real and satisfies $d\pi_M(dp) \neq 0$ on $\{p = 0\}$ is microlocally equivalent to a first order hyperbolic operator along the direction of the Hamiltonian flow. Here, $\pi_M : T^*M \to M$ is the canonical projection. Indeed, it is well known that if $dp \neq 0$ on $\{p = 0\}$, then under a 'microlocal change of variables', i.e. after conjugation by a Fourier integral operator, P is equivalent to hD_{x_1} microlocally near $\{p = 0\}$ [Zwo12, Theorem 12.3]. The crucial point in the approach from [KTZ07] is that one need only make a physical change of variables to reduce to a first order hyperbolic operator.

The main purpose of this chapter is to prove the Hörmander L^∞ estimate using the method of [KTZ07]: If $P \in \Psi^m(M)$ satisfies $\partial \overline{T}^* M \subset \mathrm{Ell}(P)$, $\sigma(P) = p$ is real, and $\{p = 0\} \cap \{d\pi_M(dp) = 0\} = \emptyset$, then

$$\|u\|_{L^\infty} \leq C h^{\frac{1-n}{2}} (\|u\|_{L^2} + h^{-1} \|Pu\|_{L^2}).$$

This result is stated and proved below in Theorem 5.3.1.

5.1 Basic Estimates

In this section, we review some basic estimates that will be useful in proving our sup-norm estimates. First, we consider a first order hyperbolic problem and review the standard energy estimate. In what follows, for $(x, y) \in \mathbb{R}^{n_1} \times \mathbb{R}^{n_2}$, with coordinates $(x, y, \xi, \eta) \in T^*\mathbb{R}^{n_1} \times \mathbb{R}^{n_2}$, and $a \in C^\infty(\mathbb{R}^{n_1}_x; S^m(T^*\mathbb{R}^{n_2}_{(y,\eta)}))$, we use the notation

$$A(x, y, hD_y) := Op_h^L(a(x, y, \eta)),$$

© The Author(s), under exclusive license to Springer Nature Switzerland AG 2023
Y. Canzani and J. Galkowski, *Geodesic Beams in Eigenfunction Analysis*, Synthesis
Lectures on Mathematics & Statistics, https://doi.org/10.1007/978-3-031-31586-2_5

with Op_h^L as in Definition 3.1.6.

Lemma 5.1.1 *Let* $\Upsilon \in C^\infty(\mathbb{R}_t; \Psi^1(\mathbb{R}^{d-1}))$ *with real principal symbol. Then there is* $C > 0$ *such that for all* $0 < T < 1$ *and* $0 < h < 1$,

$$\|v(0)\|_{L^2(\mathbb{R}^{d-1})}$$
$$\leq e^{CT}\left(\frac{\|v\|_{L^2(0,T)\times\mathbb{R}_{d-1}}}{\sqrt{T}} + \frac{\sqrt{2T}}{h}\|(hD_t - \Upsilon(t,x,hD_x))v\|_{L^2((0,T)\times\mathbb{R}^{d-1})}\right).$$

Proof First, suppose that $v(t)$ solves

$$(hD_t - \Upsilon(t,x,hD_x))v = 0.$$

Then, with

$$E(t) := \frac{1}{2}\|v(t)\|^2_{L^2(\mathbb{R}^{d-1})},$$

we have

$$hD_t E(t) = 2i\Im\langle hD_t v(t), v(t)\rangle_{L^2_x}$$
$$= 2i\Im\langle \Upsilon(t,x,hD_x)v(t), v(t)\rangle_{L^2_x}$$
$$= \langle(\Upsilon(t,x,hD_x) - [\Upsilon(t,x,hD_x)]^*)v(t), v(t)\rangle_{L^2_x}.$$

Now, since Υ has real principal symbol, $\sigma(\Upsilon(t,x,h,D_x)) = \sigma([\Upsilon(t,x,h,D_x)]^*)$, and hence

$$\Upsilon(t,x,hD_x) - [\Upsilon(t,x,hD_x)]^* \in C^\infty(\mathbb{R}_t; h\Psi^0(\mathbb{R}^{d-1})).$$

In particular, for $-1 \leq t \leq 1$,

$$\partial_t E(t) \leq C\|v(t)\|^2_{L^2} = CE(t).$$

Thus, by Grönwall's inequality, for all $-1 \leq t \leq 1$,

$$\|v(t)\|^2_{L^2_x} = 2E(t) \leq 2e^{C|t|}E(0) = e^{C|t|}\|v(0)\|^2_{L^2_x}.$$

Now, let $U(t) : L^2(\mathbb{R}^{d-1}) \to L^2(\mathbb{R}^{d-1})$ denote the solution to

$$(hD_t - \Upsilon(t,x,hD_x))U(t)u_0 = 0, \qquad U(0)u_0 = u_0,$$

Then by the preceding analysis, we have for $|s| \leq 1$,

$$\|U(s)\|_{L^2(\mathbb{R}^{d-1})\to L^2(\mathbb{R}^{d-1})} \leq e^{C|s|}. \tag{5.1}$$

Now, suppose v solves

$$(hD_t - \Upsilon(t,x,hD_x))v = f.$$

Let $\chi \in C_c^\infty((-1,1);[0,1])$ with $\chi \equiv 1$ near 0, and $|\chi'| \leq 2$ and put $\chi_T(t) = \chi(T^{-1}t)$. Then,

$$(hD_t - \Upsilon(t,x,hD_x))\chi_T v = \chi_T f - ihT^{-1}\chi'(T^{-1}t)v(t) =: f_T,$$

and therefore by Duhamel's formula, using that $\chi_T(0) = 1$, and $\operatorname{supp}\chi_T \subset (-T,T)$ we have

$$v(0) = -\frac{i}{h}\int_0^T U(-s)f_T(s)ds.$$

In particular, applying (5.1) together with the Cauchy–Schwarz inequality and using that $|T| \leq 1$, we obtain

$$\begin{aligned}
\|v(0)\|_{L^2}^2 &\leq e^{CT}h^{-2}\Big\|\int_0^T U(-s)f_T(s)ds\Big\|_{L_x^2}^2 \\
&\leq e^{CT}h^{-2}T\int_0^T \|U(-s)f_T(s)\|_{L_x^2}^2 ds \\
&\leq e^{CT}h^{-2}T\int_0^T \|f_T(s)\|_{L_x^2}^2 ds \\
&\leq e^{CT}h^{-2}T\|f_T\|_{L^2((0,T)\times\mathbb{R}^{d-1})}^2.
\end{aligned} \tag{5.2}$$

Finally, observe that

$$\|f_T\|_{L^2((0,T)\times\mathbb{R}^{d-1})} \leq \|f\|_{L^2((0,T)\times\mathbb{R}^{d-1})} + 2hT^{-1}\|v\|_{L^2((0,T)\times\mathbb{R}^{d-1})},$$

and hence the Lemma follows from (5.2). $\qquad\qquad\square$

The next technical estimate that we will need is a version of the Sobolev embedding into L^∞.

Lemma 5.1.2 *Let $k > d/2$. Then there is $C > 0$ such that for all $u \in \mathcal{D}'(\mathbb{R}^d)$ and $\{\Xi_i\}_{i=1}^d$, we have*

$$\|u\|_{L^\infty} \leq Ch^{-\frac{d}{2}}\|u\|_{L^2}^{1-\frac{d}{2k}}\Big(\sum_{i=1}^d \|(hD_{x_i} - \Xi_i)^k u\|_{L^2}^2\Big)^{\frac{d}{4k}}.$$

Proof First, observe that it is enough to prove the Lemma with $\Xi_i = 0, i = 1, \ldots, d$. Indeed, if we put $\Xi := (\Xi_1, \ldots, \Xi_d)$ and

$$v := e^{-\frac{i}{h}\langle x, \Xi\rangle}u,$$

then

$$\|u\|_{L^\infty} = \|v\|_{L^\infty}, \qquad \|u\|_{L^2} = \|v\|_{L^2}, \qquad \|(hD_{x_i} - \Xi_i)^k u\|_{L^2} = \|(hD_{x_i})^k v\|_{L^2}.$$

We assume from now on that $\Xi \equiv 0$. Without loss of generality, we also assume $\|u\|_{L^2} \neq 0$. Let $\psi \in C_c^\infty(\mathbb{R}^d)$ with $\psi \equiv 1$ on $|x| \leq 1$ and for $\varepsilon > 0$, put $\psi_\varepsilon(x) = \psi(\varepsilon^{-1}x)$. Finally, let

$$f^2(\xi) := \sum_{i=1}^d \xi_i^{2k}$$

Then, applying the Cauchy–Schwarz inequality

$$|u(x)| = \left| \frac{1}{(2\pi h)^d} \int e^{\frac{i}{h}\langle x,\xi\rangle} \psi_\varepsilon(\xi)\mathcal{F}_h(u)(\xi)d\xi + \frac{1}{(2\pi h)^d} \int e^{\frac{i}{h}\langle x,\xi\rangle}(1 - \psi_\varepsilon(\xi)\mathcal{F}_h(u)(\xi)d\xi \right|$$

$$\leq \frac{1}{(2\pi h)^d} \int |\psi_\varepsilon(\xi)\mathcal{F}_h(u)(\xi)|d\xi + \frac{1}{(2\pi h)^d} \int \frac{|1 - \psi_\varepsilon(\xi)|}{f(\xi)} f(\xi)\mathcal{F}_h(u)(\xi)|d\xi$$

$$\leq C h^{-d}(\|\psi_\varepsilon(\xi)\|_{L^2}\|\mathcal{F}_h(u)\|_{L^2} + \left\|\tfrac{1}{f}(\xi)(1 - \psi_\varepsilon)\right\|_{L^2}\|f(\xi)\mathcal{F}_h(u)\|_{L^2}).$$

$$\leq C h^{-\frac{d}{2}}\varepsilon^{\frac{d}{2}}\left[\|u\|_{L^2} + \varepsilon^{-k}\left(\sum_{i=1}^d \|(hD_{x_i})^k u\|_{L^2}^2 \right)^{\frac{1}{2}} \right]$$

where we have used that

$$\|\psi_\varepsilon\|_{L^2} = \varepsilon^{\frac{d}{2}}\|\psi\|_{L^2}, \qquad \left\|\tfrac{1}{f}(1 - \psi_\varepsilon)\right\|_{L^2} \leq \varepsilon^{\frac{d}{2}-k}\left\|\tfrac{1}{f}(1 - \psi)\right\|_{L^2}$$

and

$$(2\pi h)^{-n}\|f\mathcal{F}_h(u)\|_{L^2}^2 = (2\pi h)^{-n} \int \sum_{i=1}^d \xi_i^{2k}|\mathcal{F}_h(u)|^2(\xi)d\xi = \sum_{i=1}^d \|(hD_{x_i})^k u\|_{L^2}^2.$$

Now, choosing

$$\varepsilon = \|u\|_{L^2}^{-1/k}\left(\sum_{i=1}^d \|(hD_{x_i})^k u\|_{L^2}^2 \right)^{\frac{1}{2k}},$$

we obtain

$$\|u\|_{L^\infty} \leq C h^{-\frac{d}{2}}\|u\|^{1-\frac{d}{2k}}\left(\sum_{i=1}^d \|(hD_{x_i})^k u\|_{L^2}^2 \right)^{\frac{d}{4k}}.$$

$\square$

5.2 Factorization

In this section, we show how to factor an operator P near a point (x_0, ξ_0) where its symbol vanishes: $p(x_0, \xi_0) = 0$. The crucial fact which makes it possible to study L^∞ norms is that no microlocal changes of variables are necessary. Throughout this section, we write $x = (x_1, x') \in \mathbb{R} \times \mathbb{R}^{n-1}$ with dual variable $\xi = (\xi_1, \xi')$.

Lemma 5.2.1 *Let $P \in \Psi^m(\mathbb{R}^n)$ with symbol p such that $p(x_0, \xi_0) = 0$ and $|\partial_{\xi_1} p(x_0, \xi_0)| > c > 0$. Then there is a neighborhood, U, of (x_0, ξ_0), $E \in \Psi^{\mathrm{comp}}$ with*

$$\inf_{(x,\xi)\in\overline{U}} |\sigma(E)(x, \xi)| > \frac{1}{2}|\partial_{\xi_1} p(x_0, \xi_0)|,$$

$a \in C^\infty(\mathbb{R}; C_c^\infty(T^*\mathbb{R}^{d-1}))$, and $R \in \Psi^{\mathrm{comp}}$ such that for any $B \in \Psi^{\mathrm{comp}}$ with $WF_h(B) \subset U$,

$$PB = E(hD_{x_1} - a(x_1, x', hD_{x'}))B + hRB + O(h^\infty)_{\Psi-\infty}.$$

Proof First, observe that there is a neighborhood U_0 of (x_0, ξ_0) such that

$$\inf_{(x,\xi)\in V} |\partial_{\xi_1} p(x, \xi)| > \frac{2}{3}|\partial_{\xi_1} p(x_0, \xi_0)|.$$

Next, by the implicit function theorem, there is a neighborhood W_0' of (x_0, ξ_0') and $\tilde{a} \in C^\infty(W_0')$ such that $p(x, \tilde{a}(x, \xi'), \xi') = 0$ for $(x, \xi') \in W_0'$. In particular, since $\partial_{\xi_1} p(x_0, \xi_0) \neq 0$ on U_0, there is $U_1 \Subset U_0$ a neighborhood of (x_0, ξ_0) such that, with

$$U_2 := \{(x, \xi_1, \xi') : \xi_1 \in \mathbb{R}, (x, \xi') \in W_0'\} \cap U_1,$$

we have

$$e(x, \xi) = \frac{p(x, \xi)}{\xi_1 - \tilde{a}(x, \xi')} = \frac{p(x, \xi) - p(x, \tilde{a}(x, \xi'), \xi')}{\xi_1 - \tilde{a}(x, \xi')} \in C^\infty(U_2),$$

and $|e(x, \xi)| > \frac{1}{2}|\partial_{\xi_1} p(x_0, \xi_0)|$ on U_2.

Let $W_1' \Subset W_0'$ be a neighborhood of (x_0, ξ_0'), $U_3 \Subset U_2$ a neighborhood of (x_0, ξ_0), and

$$U_4 := \{(x, \xi_1, \xi') : \xi_1 \in \mathbb{R}, (x, \xi') \in W_1'\} \cap U_3,$$

Next, let $\chi \in C_c^\infty(U_1)$ with $\chi \equiv 1$ on U_4, $\psi \in C_c^\infty(W_0')$ with $\psi \equiv 1$ on W_1', and put $E = Op_h(e\chi)$, $a = \tilde{a}\psi$. Then, for $(x, \xi) \in U_4$,

$$\sigma(E(hD_{x_1} - a(x_1, x', hD_{x'}))) = e(x, \xi)(\xi_1 - \tilde{a}(x, \xi')) = p(x, \xi).$$

Finally, let $U \Subset U_4$ be a neighborhood of (x_0, ξ_0) and let B' with $WF_h(B') \subset U_4$ and $WF_h(I - B') \cap U = \emptyset$. Then, we have

$$\sigma(E(hD_{x_1} - a(x_1, x', hD_{x'}))B') = \sigma(PB'),$$

and hence

$$E(hD_{x_1} - a(x_1, x', hD_{x'}))B' - PB' = hR,$$

for $R \in \Psi^{\mathrm{comp}}$. In particular, for B with $WF_h(B) \subset U$,

$$PB = PB'B + O(h^\infty)_{\Psi-\infty}$$
$$= E(hD_{x_1} - a(x_1, x', hD_{x'}))B'B + hRB + O(h^\infty)_{\Psi-\infty}$$
$$= E(hD_{x_1} - a(x_1, x', hD_{x'}))B + hRB + O(h^\infty)_{\Psi-\infty},$$

as claimed. $\square$

5.3   The Hörmander L^∞ Estimate

We are now in a position to prove the Hörmander L^∞ estimate.

Theorem 5.3.1 (L^∞ bound) *Let M be a smooth compact manifold, $s > \frac{n}{2}$ and suppose that $P \in \Psi^m$ with symbol $p := \sigma(P)$ satisfies*

$$\partial \overline{T}^*M \subset \mathrm{Ell}(P), \qquad \inf_{T^*M} |p| + |d_\xi p| > c > 0.$$

Then there is $C > 0$ such that for all $u \in \mathcal{D}'(M)$,

$$\|u\|_{L^\infty} \leq Ch^{\frac{1-n}{2}}(\|u\|_{L^2} + h^{-1}\|Pu\|_{H_h^{s-m}(M)}).$$

Proof To prove the theorem we will use a microlocal partition of unity to work near a point $\rho \in \overline{T}^*M$.

We first consider the simpler case of $\rho \in \mathrm{Ell}(P)$. Let $\rho \in \mathrm{Ell}(P)$. Then, there is a neighborhood $U_\rho \subset \mathrm{Ell}(P)$ of ρ. Let $V_\rho \Subset U_\rho$ be a neighborhood of ρ and $X_\rho \in \Psi^0(M)$ with $\mathrm{WF_h}(X_\rho) \subset U_\rho$, $\mathrm{WF_h}(I - X_\rho) \cap V_\rho = \emptyset$. Then, by Lemma 3.3.2, there is $E_\rho \in \Psi^0$ such that

$$X_\rho = E_\rho P + O(h^\infty)_{\Psi-\infty}.$$

In particular, for all B_ρ with $\mathrm{WF_h}(B_\rho) \subset V_\rho$, and $N > 0$ we have

$$\|B_\rho u\|_{H_h^s} \leq C\|Pu\|_{H_h^{s-m}} + C_N h^N \|u\|_{H_h^{-N}},$$

and hence, by Lemma 5.1.2,

$$\|B_\rho u\|_{L^\infty} \leq Ch^{-\frac{d}{2}}\|Pu\|_{H_h^{s-m}} + C_N h^N \|u\|_{H_h^{-N}} \tag{5.3}$$

Next, we consider the case of $\rho \notin \mathrm{Ell}(P)$. Then, by assumption

$$d_\xi p(\rho) \neq 0.$$

Therefore, we may choose coordinates, x on M in a neighborhood of $\pi_M(\rho)$ such that $\partial_{\xi_1} p(\rho) \neq 0$ and hence Lemma 5.2.1 applies. Thus, there is a neighborhood, V_ρ of ρ, $E \in$

Ψ^{comp} with $\overline{V_\rho} \subset \mathrm{Ell}(E_\rho)$, $a \in C^\infty(\mathbb{R}; C_c^\infty(T^*\mathbb{R}^{d-1}))$, and $R \in \Psi^{\mathrm{comp}}$ such that for B_ρ with $\mathrm{WF_h}(B_\rho) \subset V_\rho$, we have

$$P B_\rho = E(hD_{x_1} - a(x_1, x', hD_{x'})) B_\rho + hR B_\rho + O(h^\infty)_{\Psi-\infty}.$$

Let $X \in \Psi^{\mathrm{comp}}$ with $\mathrm{WF_h}(I - X) \cap V_\rho = \emptyset$, and $\mathrm{WF_h}(X) \subset \mathrm{Ell}(E)$. Then, by Lemma 3.3.2 there is $W \in \Psi^{\mathrm{comp}}$ such that

$$W E = X + O(h^\infty)_{\Psi-\infty}.$$

In particular, since $\mathrm{WF_h}(I - X) \cap \mathrm{WF_h}(B_\rho) = \emptyset$,

$$\begin{aligned}
W P B_\rho &= X(hD_{x_1} - a(x_1, x', hD_{x'})) B_\rho + hX R B_\rho + O(h^\infty)_{\Psi-\infty} \\
&= (hD_{x_1} - a(x_1, x', hD_{x'})) B_\rho + hR B_\rho + O(h^\infty)_{\Psi-\infty}.
\end{aligned}$$

Therefore, applying Lemma 5.1.1 with x_1 playing the role of t, we have for any x_1,

$$\begin{aligned}
\|[B_\rho u](x_1)\|_{L^2_{x'}} &\leq C(\|B_\rho u\|_{L^2} + h^{-1}\|W P B_\rho u\|_{L^2} + O(h^\infty)\|u\|_{H_h^{-N}}) \\
&\leq C(\|u\|_{L^2} + h^{-1}\|Pu\|_{H_h^{s-m}}).
\end{aligned}$$

Similarly, for any k,

$$\begin{aligned}
\left\|[(hD_{x'})^k B_\rho u](x_1)\right\|_{L^2_{x'}} &\leq C(\|(hD_{x'})^k B_\rho u\|_{L^2} \\
&\quad + h^{-1}\|W P (hD_{x'})^k B_\rho u\|_{L^2} + O(h^\infty)\|u\|_{H_h^{-N}}) \\
&\leq C(\|u\|_{L^2} + h^{-1}\|Pu\|_{H_h^{s-m}}).
\end{aligned}$$

Thus, by Lemma 5.1.2,

$$\|[B_\rho u](x_1)\|_{L^\infty_{x'}} \leq Ch^{\frac{1-d}{2}}(\|u\|_{L^2} + h^{-1}\|Pu\|_{L^2}),$$

and, since x_1 was arbitrary, this implies

$$\|[B_\rho u]\|_{L^\infty} \leq Ch^{\frac{1-d}{2}}(\|u\|_{L^2} + h^{-1}\|Pu\|_{H_h^{s-m}}). \tag{5.4}$$

By (5.3) and (5.4), for all $\rho \in \overline{T}^*M$, there is a neighborhood U_ρ of ρ such that for $B_\rho \in \Psi^0(M)$ with $\mathrm{WF_h}(B_\rho) \subset U_\rho$,

$$\|B_\rho u\|_{L^\infty} \leq Ch^{\frac{1-d}{2}}(\|u\|_{L^2} + h^{-1}\|Pu\|_{H_h^{s-m}}). \tag{5.5}$$

In particular, by compactness of $\overline{T}^*M$, there are $\{\rho_i\}_{i=1}^N$ such that $\overline{T}^*M \subset \bigcup_{i=1}^N U_{\rho_i}$. Let $\{\chi_i\}_{i=1}^N \subset S^0(T^*M)$ be a partition of unity subordinate to $\{U_{\rho_i}\}_{i=1}^N$. Then, since $\mathrm{WF_h}(Op_h(\chi_i)) \subset U_{\rho_i}$,

$$\|u\|_{L^\infty} = \left\| \sum_{i=1}^{N} Op_h(\chi_i)u \right\|_{L^\infty}$$

$$\leq \sum_{i=1}^{N} \|Op_h(\chi_i)u\|_{L^\infty}$$

$$\leq Ch^{\frac{1-d}{2}}(\|u\|_{L^2} + h^{-1}\|Pu\|_{H_h^{s-m}}),$$

where we have used (5.5) in the last line. $\qquad\square$

Geodesic Beam Tools **6**

In this chapter, we discuss the tools at the heart of the geodesic beam analysis. That is, the construction of the beams themselves, as well as the corresponding improved estimates. Given a function, u which is a quasimode for $(-h^2\Delta_g - 1)$, the geodesic beams take the form

$$u_\chi = Op_h(\chi)u,$$

where χ is a localizer to a geodesic tube of length ~ 1 and radius $\sim R(h)$ along a geodesic that nearly commutes with the Laplacian near a point. See Definition 6.1.2 for the precise formulation.

6.1 Basic Geodesic Beam Estimates

In this section, we revisit the proof of Hörmander's L^∞ estimate via the Koch–Tataru–Zworski method to give improved estimates on the L^∞ norms of functions when they are localized using cutoffs that are adapted to the Laplacian and supported near a geodesic segment.

To state this estimate, we first define for $\rho \in S^*M$,

$$\Lambda_\rho^\tau := \{\varphi_t(\rho) : t \in [-\tau, \tau]\}. \tag{6.1}$$

Lemma 6.1.1 (Beam estimate) *There are $C_n, \tau_0, \varepsilon_0, R_0 > 0$ such that for all $\rho_0 = (x_0, \xi_0) \in S^*M, 0 \leq \delta < \frac{1}{2}, 0 < \tau < \tau_0, 0 < \varepsilon < \varepsilon_0, R(h) : (0,1) \to (0, R_0)$ with $R(h) \geq h^\delta$, and $X \in \Psi_\delta^{\mathrm{comp}}(M)$ satisfying*

$$\mathrm{MS}_h(X) \subset \{\rho : d(\Lambda_{\rho_0}^{\tau+\varepsilon}, \rho) < R(h)\}, \qquad \mathrm{WF}_h([-h^2\Delta_g, X]) \cap B(x_0, \tau) = \emptyset, \tag{6.2}$$

© The Author(s), under exclusive license to Springer Nature Switzerland AG 2023
Y. Canzani and J. Galkowski, *Geodesic Beams in Eigenfunction Analysis*, Synthesis
Lectures on Mathematics & Statistics, https://doi.org/10.1007/978-3-031-31586-2_6

we have that for all $N > 0$ there is $C_N > 0$ such that for $0 < h < 1$ and $u \in \mathcal{D}'(M)$,

$$\|Xu\|_{L^\infty(B(x_0,\tau/6))} \le \frac{C_n}{\tau^{1/2}} h^{\frac{1-n}{2}} R(h)^{\frac{n-1}{2}} \left(\|Xu\|_{L^2} + h^{-1}\|X(-h^2\Delta_g - 1)u\|_{L^2} + C_N h^N \|u\|_{L^2} \right).$$

Definition 6.1.2 (*Geodesic beam*) If $\|(-h^2\Delta_g - 1)u\|_{L^2} = o(h)\|u\|_{L^2}$, and X satisfies (6.2), then we call Xu a *geodesic beam*.

The key fact, which we prove in Lemma 6.2.7, is that a quasimode, u, can be decomposed into geodesic beams locally in phase space.

Proof Let $P = -h^2\Delta_g - 1$ with symbol $p = |\xi|_g^2 - 1$. For all $\rho \in S^*M$, we may choose coordinates on M such that $\partial_{\xi_1} p(\rho) = |\partial_\xi p(\rho)|$. Therefore, by Lemma 5.2.1 there is a neighborhood, U_ρ, of ρ, $E, R \in \Psi^{\mathrm{comp}}$, and $a \in C^\infty(\mathbb{R}; C_c^\infty(T^*\mathbb{R}^{d-1}))$ such that for any $B \in \Psi^{\mathrm{comp}}$ with $\mathrm{WF}_h \subset U_\rho$,

$$PB = E(hD_{x_1} - a(x_1, x', hD_{x'}))B + hRB + O(h^\infty)_{\Psi^{-\infty}}, \tag{6.3}$$

and

$$|\sigma(E)(\rho_1)| > \frac{1}{2}|\partial_\xi p(\rho)|, \qquad \rho_1 \in U_\rho.$$

Shrinking U_ρ if necessary, we may assume that

$$\inf_{U_\rho} |\partial_\xi p| > \frac{1}{2}|\partial_\xi p(\rho)|.$$

By compactness, there are $\{\rho_i\}_{i=1}^N$ such that $S^*M \subset \bigcup_{i=1}^N U_{\rho_i}$ and there is $\tilde{R}_0 > 0$ such that for every $\rho_0 \in S^*M$, there is i such that $B(\rho, \tilde{R}_0) \subset U_{\rho_i}$.

In particular, we may choose $\tau_0, \varepsilon_0, R_0 > 0$ such that for all $\rho \in S^*M$ there is i such that $\Lambda_{\rho_0}^{\tau_0+\varepsilon_0}(R_0) \subset U_{\rho_i}$. We can now assume that $\mathrm{WF}_h \subset U_{\rho_i}$ for some i and hence that we may use coordinates where (6.3) holds.

Let $q \in C^\infty(\mathbb{R}_{x_1}; C_c^\infty(T^*\mathbb{R}^{n-1}))$, put $Q = q(x, hD_{x'})$, and define

$$\begin{aligned} f &:= (hD_{x_1} - a(x, hD_{x'}))QXu \\ &= E^{-1}QXPu + E^{-1}[P, QX]u + hE^{-1}RQXu + O(h^\infty)_{\Psi^{-\infty}}u, \end{aligned} \tag{6.4}$$

where $E^{-1} \in \Psi^{\mathrm{comp}}$ such that $\mathrm{WF}_h(E^{-1}E - I) \cap \overline{U} = \emptyset$.

Let $S_T := (-2T, 2T)_{x_1} \times \mathbb{R}^{n-1}$. Note that for $\mathcal{A} \subset \Psi_\delta^{\mathrm{comp}}$ bounded there is $h_0 > 0$ such that for all $A \in \mathcal{A}$ with $|\sigma(A)| \le 1$ and $\chi, \tilde{\chi} \in C_c^\infty(-2T, 2T)$ with $\chi \equiv 1$ on $(-T, T)$ and $\tilde{\chi} \equiv 1$ on $\mathrm{supp}\,\chi$, and all $0 < h < h_0$,

$$\|Au\|_{L^2((-T,T)_{x_1} \times \mathbb{R}^{n-1})} \le \|\chi Au\|_{L^2} \le \|\chi A\tilde{\chi}u\|_{L^2} + C_N h^N \|u\|_{L^2}$$

$$\le 2\|u\|_{L^2(S_T)} + C_N h^N \|u\|_{L^2}.$$

Then, by Lemma 5.1.1 and (6.4), we have for $|x_1| < T/2$ and $0 < T < \tau/2$

$$\|[QXu](x_1, \cdot)\|_{L^2_{x'}} \leq \frac{e^{CT}}{\sqrt{T}} \|QXu\|_{L^2(S_T)}$$
$$+ \frac{e^{CT}\sqrt{2T}}{h}\Big[\|QXPu\|_{L^2(S_T)} + \|[P,Q]Xu\|_{L^2(S_T)} + \|Q[P,X]u\|_{L^2(S_T)}\Big]$$
$$+ Ce^{CT}\sqrt{T}\|QXu\|_{(L^2(S_T))} + C_N h^N e^{CT}\sqrt{T}\|u\|_{L^2}.$$

Now, since $0 < T \leq \tau/3$, $\mathrm{MS_h}([P,X]) \cap S_T = \emptyset$, and hence

$$\|[QXu](x_1, \cdot)\|_{L^2_{x'}} \leq \frac{e^{CT}}{\sqrt{T}}\|QXu\|_{L^2(S_T)} + \frac{e^{CT}\sqrt{2T}}{h}\Big[\|QXPu\|_{L^2(S_T)} + \|[P,Q]Xu\|_{L^2(S_T)}\Big]$$
$$+ Ce^{CT}\sqrt{T}\|QXu\|_{(L^2(S_T))} + C_N h^N \|u\|_{L^2}. \tag{6.5}$$

Next, we choose q judiciously to obtain our improved L^∞ estimate. Once again, using that $\partial_{\xi_1} p > 0$, we have

$$\partial_t(x_1(\varphi_t(\rho_0))) > 0$$

and in particular, we can reparametrize the geodesics so by x_1. i.e. there is $t(x_1)$ such that

$$x_1(\varphi_{t(r)}(\rho_0)) = r.$$

Putting, $(x'(x_1), \xi(x_1)) := (x'(\varphi_{t(x_1)}(\rho_0)), \xi(\varphi_{t(x_1)}))$ we have that

$$x_1 \mapsto (x_1, x'(x_1), \xi(x_1))$$

parametrizes the geodesic through ρ_0. Now, define q_i by

$$\tilde{q}_i = (\xi_i - \xi_i(x_1)), \qquad Q_i = [\tilde{q}_i(x, hD_{x'})]^k, \qquad q_i = \sigma(Q_i) = (\xi_i - \xi_i(x_1))^k.$$

Then, $q_i = (\xi_i - \xi_i(x_1))^k$ vanishes to order k on γ. Moreover, since

$$H_p q_i = k(\xi' - \xi'(x_1))^{k-1} H_p(\xi' - \xi'(x_1))$$

and $\xi' - \xi'(x_1) \equiv 0$ on γ, $H_p q_i$ also vanishes to order k along γ.

We now use this vanishing to improve estimates on $Q_i X$. We have

$$Q_i X = Q_i Op_h(\tilde{\chi})X + O(h^\infty)_{\Psi-\infty}$$

where $\tilde{\chi}$ is supported in $\{\rho : d(\Lambda^{\tau+\varepsilon}_{\rho_0}, \rho) < R(h)\}$ and has $\tilde{\chi} \equiv 1$ on $\mathrm{MS_h}(X)$. Then, since q_i is controlled by the distance to γ to the power k, we have

$$\sup |\sigma(Q_i Op_h(\tilde{\chi}))| \leq R(h)^k,$$

and $Q_i Op_h(\tilde{\chi}) \in R(h)^k S_\delta^{\text{comp}}$. Therefore, we obtain

$$\|Q_i Op_h(\tilde{\chi})\|_{L^2 \to L^2} \le R(h)^k (1 + O(h^{1-2\delta}))).$$

In particular,

$$\|Q_i Xu\|_{L^2} \le R(h)^k (1 + O(h^{1-2\delta}))\|Xu\|_{L^2} + C_N h^N \|u\|_{L^2},$$
$$\|Q_i XPu\|_{L^2} \le R(h)^k (1 + O(h^{1-2\delta}))\|XPu\|_{L^2} + C_N h^N \|u\|_{L^2}. \tag{6.6}$$

Similarly, since $H_p q_i$ vanishes to order k on γ,

$$\|[P, Q_i]Xu\|_{L^2} \le ChR(h)^k (1 + O(h^{1-2\delta}))\|Xu\|_{L^2} + C_N h^N \|u\|_{L^2}. \tag{6.7}$$

Using (6.6) and (6.7) in (6.5) with $Q = Q_i$ we obtain

$$\|[QXu](x_1, \cdot)\|_{L^2_{x'}} \le e^{CT} \left(\frac{1}{\sqrt{T}} + C\sqrt{T}\right) R(h)^k (1 + Ch^{1-2\delta})\|Xu\|_{L^2}$$
$$+ \frac{e^{CT}\sqrt{2T}}{h} \left[R(h)^k (1 + Ch^{1-2\delta})\|XPu\|_{L^2}\right] + C_N h^N \|u\|_{L^2}.$$

Shrinking τ_0 if necessary, so that $e^{C\tau_0}(1 + C\sqrt{\tau_0}) < 2$, choosing h_0 small enough, and using that $Q_i = (hD_{x_i} - \xi_i(x_1))^k$, we have for any i, k

$$\|[(hD_{x_i} - \xi_i(0))^k Xu](x_1, \cdot)\|_{L^2} \le \frac{4R(h)^k}{\sqrt{T}}(\|Op_h(\chi)u\|_{L^2} + h^{-1}\|Op_h(\chi)Pu\|_{L^2})$$
$$+ C_N h^N \|u\|_{L^2}. \tag{6.8}$$

Using (6.8) with $k = 0$ and $k = n - 1$, $i = 2, \ldots n - 1$, and applying Lemma 5.1.2 with $d = n - 1$,

$$\|Xu\|_{L^\infty(B(0,\tau/6))} \le \frac{C_n}{\sqrt{\tau}} h^{\frac{1-n}{2}} R(h)^{\frac{n-1}{2}} (\|Xu\|_{L^2} + h^{-1}\|XPu\|_{L^2}) + C_N h^N \|u\|_{L^2}$$

$$\square$$

Next, we present the analogue of Lemma 6.1.1 with the point x_0 replaced by a submanifold H.

Lemma 6.1.3 *Let $1 \le k \le n$. There are $C_{n,k} > 0$ such that for all $H \subset M$ smooth embedded submanifold of codimension k there are $\tau_0, \varepsilon_0, R_0 > 0$ such that for all $\rho_0 = (x_0, \xi_0) \in SN^*H$, $0 \le \delta < \frac{1}{2}$, $0 < \tau < \tau_0$, $0 < \varepsilon < \varepsilon_0$, $R(h) : (0, 1) \to (0, R_0)$ with $R(h) \ge h^\delta$, and $X \in \Psi_\delta^{\text{comp}}(M)$ satisfying*

$$\mathrm{MS}_h(X) \subset \{\rho : d(\Lambda_{\rho_0}^{\tau+\varepsilon}, \rho) < R(h)\}, \qquad \mathrm{WF}_h([-h^2 \Delta_g, X]) \cap B(x_0, \tau) = \emptyset,$$

we have that for all $N > 0$ there is $C_N > 0$ such that for $0 < h < 1$, $w \in C_c^\infty(H)$ with $|w| \leq 1$, $u \in \mathcal{D}'(M)$,

$$\|wXu\|_{L^1(H)} \leq \frac{C_{n,k}}{\tau^{1/2}} h^{\frac{1-k}{2}} R(h)^{\frac{n-1}{2}} \left(\|Xu\|_{L^2} + h^{-1}\|X(-h^2\Delta_g - 1)u\|_{L^2} + C_N h^N \|u\|_{L^2} \right).$$

Remark 6.1.4 It is clear from the proof below that a version of Lemma 6.1.3 with some uniformity in the submanifold H holds (see also [CG21, Lemma 3.7]). This uniformity involves assumptions on closeness of the submanifolds in C^k for some large enough k.

Proof Let $P = -h^2\Delta_g - 1$ with symbol $p = |\xi|_g^2 - 1$. For all $\rho \in SN^*H$, we may choose coordinates $x = (x_1, x') \in \mathbb{R} \times \mathbb{R}^{n-1}$ with $x' = (x'', y) \in \mathbb{R}^{k-1} \times \mathbb{R}^{n-k}$ on M such that $\partial_{\xi_1} p(\rho) = |\partial_\xi p(\rho)|$ and $|(x_1, x'')| = d(x, H)$. Therefore, by Lemma 5.2.1 there is a neighborhood, U_ρ, of ρ, $E, R \in \Psi^{\mathrm{comp}}$, and $a \in C^\infty(\mathbb{R}; C_c^\infty(T^*\mathbb{R}^{d-1}))$ such that for any $B \in \Psi^{\mathrm{comp}}$ with $\mathrm{WF}_h(B) \subset U_\rho$,

$$PB = E(hD_{x_1} - a(x_1, x', hD_{x'}))B + hRB + O(h^\infty)_{\Psi^{-\infty}}, \tag{6.9}$$

and

$$|\sigma(E)(\rho_1)| > \frac{1}{2}|\partial_\xi p(\rho)|, \qquad \rho_1 \in U_\rho.$$

Shrinking U_ρ if necessary, we may assume that

$$\inf_{U_\rho} |\partial_\xi p| > \frac{1}{2}|\partial_\xi p(\rho)|.$$

By compactness, there are $\{\rho_i\}_{i=1}^N$ such that $SN^*H \subset \bigcup_{i=1}^N U_{\rho_i}$ and there is $\tilde{R}_0 > 0$ such that for every $\rho_0 \in SN^*H$, there is i such that $B(\rho, \tilde{R}_0) \subset U_{\rho_i}$.

In particular, we may choose $\tau_0, \varepsilon_0, R_0 > 0$ such that for all $\rho \in SN^*H$ there is i such that $\Lambda_{\rho_0}^{\tau_0+\varepsilon_0}(R_0) \subset U_{\rho_i}$. We can now assume that $\mathrm{WF}_h(X) \subset U_{\rho_i}$ for some i and hence that we may use coordinates where (6.9) holds. Without loss of generality, we assume that $\rho = (0, 0, 0, 1, 0, 0)$ and that the metric is Euclidean at $\pi_M(\rho) = (0, 0, 0) \in \mathbb{R} \times \mathbb{R}^{k-1} \times \mathbb{R}^{n-k}$.

By our choice of coordinates,

$$\Lambda_\rho^t = \{(s, 0, 0, 1, 0, 0) : |s| \leq t\}.$$

Moreover, since the metric is Euclidean at $(0, 0, 0)$, there is $T_0 < 0$ such that

$$\{q : d(\Lambda_\rho^{T_0}, q) < R(h)\} \subset \left\{ (x_1, x'', y) : |y| \leq 2R(h),\ |(x_1, x'')| \leq 1 \right\}.$$

Now, let $\psi \in C_c^\infty(\mathbb{R}^{n-k})$ with $\psi \equiv 1$ on $B(0, 1)$ and define $\psi_R(y) := \psi(R^{-1}y)$. We first claim that for any $N > 0$

$$\|\psi_{2R(h)}(Xu)|_H - Xu|_H\|_{H_h^N(H)} \leq C_N \|u\|_{H_h^{-N}(M)}. \tag{6.10}$$

Indeed, let $\chi \in C_c^\infty(\mathbb{R}^k)$ with $\chi \equiv 1$ on $B(0, 1)$ and define

$$\tilde{\psi}_{2R}(x_1, x'', y) := \chi(x_1, x'')\psi_{2R}(y).$$

Then $\tilde{\psi}_{2R(h)} \in S_\delta$ and $\mathrm{supp}(1 - \tilde{\psi}) \cap \mathrm{MS}_h(X) = \emptyset$ so that

$$Xu = \tilde{\psi}Xu + O(h^\infty)_{\Psi-\infty}u,$$

and hence (6.10) follows.

In particular, (6.10) implies

$$\|wXu\|_{L^1(H)} \leq \|w\psi_{2R(h)}Xu\|_{L^1(H)} + C_N h^N \|u\|_{L^2(M)}.$$

Now,

$$\|w\psi_{2R(h)}Xu\|_{L^1(H)} \leq \|\psi_{2R(h)}\|_{L^2(H)}\|Xu\|_{L^2(H)} \leq C_k R(h)^{\frac{n-k}{2}}\|Xu\|_{L^2(H)}.$$

Therefore, it remains to estimate $\|Xu\|_{L^2(H)}$. For this, we proceed exactly as in the proof of Lemma 6.1.1 to obtain (6.8). We then set $x_1 = 0$ and apply Lemma 5.1.2 in the x'' variables to complete the proof. $\qquad\square$

6.2   Good Covers and Partitions

When we apply the basic beam estimate, Lemma 6.1.1, in applications, we will need microlocal partitions of unity which satisfy the hypotheses there and which behave well as $h \to 0$. This section is devoted to the construction of such partitions and in particular the decomposition of a quasimode into geodesic beams.

6.2.1   Geodesic Tubes

We first build covers of subsets of S^*M by geodesic tubes that do not have too many overlaps. In order to construct our microlocal partition, we first fix a smooth hypersurface $\mathcal{S} \Subset T^*M$, such that, for any defining function $f : T^*M \to \mathbb{R}$ for $\mathcal{S}$, we have

$$|f| + |H_{|\xi|_g}f|(\rho) \geq c > 0 \text{ for } d(\rho, S^*M) < \varepsilon$$

i.e. so that the geodesic flow is transverse to $\mathcal{S}$ near S^*M. Let

$$\mathcal{H}_0 := S^*M \cap \mathcal{S}, \qquad \mathcal{H}_\varepsilon := \mathcal{S} \cap \{\rho \in T^*M : d(\rho, S^*M) < \varepsilon\}. \tag{6.11}$$

Then, $\mathcal{H}_\varepsilon$ is transverse to the geodesic flow and there is $0 < \tau_{\mathrm{inj}H} < 1$ so that the map

$$\Psi : [-\tau_{\mathrm{inj}H}, \tau_{\mathrm{inj}H}] \times \mathcal{H}_\varepsilon \to T^*M, \qquad \Psi(t, \rho) := \varphi_t(\rho), \qquad (6.12)$$

is injective.

Definition 6.2.1 (*Geodesic tube*) For $q \in \mathcal{H}_0, \tau > 0$, we define the tube through q of radius $R > 0$ and 'length' $\tau + R$ as

$$\Lambda_q^\tau(R) := \bigcup_{|t| \le \tau + R} \varphi_t(B_{\mathcal{S}}(q, R)), \quad B_{\mathcal{S}}(q, R) := \{\rho \in \mathcal{S} : d(\rho, q) \le R\}, \qquad (6.13)$$

and d is distance induced by the Sasaki metric on T^*M (See e.g. [Bla10b, Chapter 9] for a description of the Sasaki metric). Note that the tube runs along the geodesic through $q \in \mathcal{H}_0$. Similarly, for $A \subset \mathcal{H}_0$, we define $\Lambda_A^\tau(R)$ in the same way, replacing q with A in (6.13).

The following Lemma will be useful when applying Lemma 6.1.1

Lemma 6.2.2 *There are $R_0 > 0$ and $\tau_0 > 0$ such that for all $\rho_0 \in S^*M$, $0 < R < R_0$ and $0 < \tau < \tau_0$,*

$$\Lambda_{\rho_0}^\tau(R) \subset \{\rho_0 : d(\Lambda_{\rho_0}^{\tau+R}, 2R)\}.$$

Proof Let $K \subset T^*M$ compact. We first claim that there is $\delta > 0$ small enough and $C_1 > 0$ so that uniformly for $t \in [-\delta, \delta]$, and $(x_i, \xi_i) \in K$.

$$\frac{1}{3} d\big((x_1, \xi_1), (x_2, \xi_2)\big) - C_1 d\big((x_1, \xi_1), (x_2, \xi_2)\big)^2 \le d\big(\varphi_t(x_1, \xi_2), \varphi_t(x_2, \xi_1)\big)$$

$$\le \tfrac{3}{2} d\big((x_1, \xi_1), (x_2, \xi_2)\big) + C_1 d\big((x_1, \xi_1), (x_2, \xi_2)\big)^2 \qquad (6.14)$$

where d is the distance induced by the Sasaki metric.

To prove (6.14), observe that by Taylor's theorem

$$\varphi_t(x_1, \xi_1) - \varphi_t(x_2, \xi_2) = d_x\varphi_t(x_2, \xi_2)(x_1 - x_2) + d_\xi\varphi_t(x_2, \xi_2)(\xi_1 - \xi_2)$$
$$+ O_{C^\infty}\big(\sup_{q \in K} |d^2\varphi_t(q)|(|\xi_1 - \xi_2|^2 + |x_1 - x_2|^2)\big)$$

Now,

$$\varphi_t(x, \xi) = (x, \xi) + (\partial_\xi p(x, \xi)t, -\partial_x p(x, \xi)t) + O(t^2)$$

so

$$d_\xi\varphi_t(x, \xi) = (0, I) + t(\partial_\xi^2 p, -\partial_{\xi x}^2 p) + O(t^2)$$
$$d_x\varphi_t(x, \xi) = (I, 0) + t(\partial_{x\xi}^2 p, -\partial_x^2 p) + O(t^2).$$

In particular,

$$\varphi_t(x_1, \xi_1) - \varphi_t(x_2, \xi_2) = ((0, I) + O(t))(\xi_1 - \xi_2) + ((I, 0) + O(t))(x_1 - x_2)$$
$$+ O((\xi_1 - \xi_2)^2 + (x_1 - x_2)^2)$$

and choosing $\delta > 0$ small enough gives the result.

Let $\tau < \frac{\delta}{2}$ and $R < \min(\frac{1}{2C_1}, \frac{\delta}{2}$. Suppose that $\rho_0 \in \mathcal{H}_0$ let $\rho_1 \in \Lambda_{\rho_0}^\tau(R)$. Then there is $|t| < \tau + R$ and $q \in S$ such that

$$\rho_1 = \varphi_t(q), \qquad d(q, \rho_0) < R.$$

Then, by (6.14),

$$d(\rho_1, \varphi_t(\rho_0)) = d(\varphi_t(q), \varphi_t(\rho_0)) \le \tfrac{3}{2}R + C_1 R^2 \le 2R.$$

In particular,

$$\rho_1 \subset \{\rho \,:\, d(\Lambda_{\rho_0}^{\tau+R}, \rho) < 2R\},$$

and hence

$$\Lambda_{\rho_0}^\tau(R) \subset \{\rho \,:\, d(\Lambda_{\rho_0}^{\tau+R}, \rho) < 2R\}$$

as claimed. $\qquad\qquad\qquad\qquad\qquad\qquad\qquad\qquad\qquad\qquad\qquad\qquad\square$

Definition 6.2.3 (*Covers and partitions*) Let $A \subset \mathcal{H}_0, r > 0$, and $\{\rho_j(r)\}_{j=1}^{N_r} \subset A$. We say that the collection of tubes $\{\Lambda_{\rho_j}^\tau(r)\}_{j=1}^{N_h}$ is a (τ, r)-*cover* of A provided

$$\Lambda_A^\tau(\tfrac{1}{2}r) \subset \bigcup_{j=1}^{N_r} \Lambda_{\rho_j}^\tau(r).$$

In addition, for $0 \le \delta \le \frac{1}{2}$ and $R(h) \ge h^\delta$, we say that a collection $\{\chi_j\}_{j=1}^{N_h} \subset S_\delta(T^*M;$ $[0, 1])$ is a δ-*partition for* A *associated to the* $(\tau, R(h))$-*cover* if $\{\chi_j\}_{j=1}^{N_h}$ is bounded in S_δ and

(1) $\operatorname{supp} \chi_j \subset \Lambda_{\rho_j}^\tau(R(h))$,

(2) $\sum_{j=1}^{N_h} \chi_j \ge 1$ on $\Lambda_A^{\tau/2}(\frac{1}{2}R(h))$.

Although we will be able state many of our results in terms of (τ, r)-covers and δ-partitions, it will be useful in the proofs of our theorems to have covers and partitions with substantially better properties. We will construct such covers and partitions immediately following their definition.

Definition 6.2.4 (*Good covers and partitions*) Let $A \subset \mathcal{H}_0, r > 0$, and $\{\rho_j(r)\}_{j=1}^{N_r} \subset A$ be a collection of points, for some $N_r > 0$. Let $\mathfrak{D}$ be a positive integer. We say that the collection of tubes $\{\Lambda_{\rho_j}^\tau(r)\}_{j=1}^{N_r}$ is a $(\mathfrak{D}, \tau, r)$-*good cover* of $A \subset T^*M$ provided it is a (τ, r)-cover of A and there exists a partition $\{\mathcal{J}_\ell\}_{\ell=1}^{\mathfrak{D}}$ of $\{1, \ldots, N_r\}$ so that for every $\ell \in \{1, \ldots, \mathfrak{D}\}$

$$\Lambda_{\rho_j}^\tau(3r) \cap \Lambda_{\rho_i}^\tau(3r) = \emptyset, \qquad i, j \in \mathcal{J}_\ell, \qquad i \neq j.$$

In addition, for $0 \le \delta < \frac{1}{2}$ and $R(h) \ge h^\delta$, we say that a collection $\{\chi_j\}_{j=1}^{N_h} \subset S_\delta(T^*M;)$ is a δ-*good partition for* A *associated to a* $(\mathfrak{D}, \tau, R(h))$-*good cover* if there is $C > 0$ such that $\{\chi_j\}_{j=1}^{N_h}$ is bounded in $S_\delta(T^*M; [-Ch^{1-2\delta}, 1 + Ch^{1-2\delta}])$, and

- $\sum_{j=1}^{N_h} \chi_j = 1$ on $\Lambda_A^\tau(\frac{1}{2}R(h))$.
- $\operatorname{supp} \chi_j \subset \Lambda_{\rho_j}^{\tau+\varepsilon}(R(h))$,

- $\operatorname{MS_h}([P, Op_h(\chi_j)]) \cap \Lambda_{\mathcal{H}_0}^\tau(\varepsilon) = \emptyset$.

6.2.2   Construction of Good Covers and Partitions

We start by constructing a useful cover of any Riemannian manifold with bounded curvature.

Lemma 6.2.5 *Let $\tilde{M}$ be a compact Riemannian manifold. There exist $\mathfrak{D}_n > 0$, depending only on n, and $R_0 > 0$ depending only on n and a lower bound for the sectional curvature of $\tilde{M}$, so that the following holds. For $0 < r < R_0$, and $A \subset \tilde{M}$, there exist a finite collection of points $\{x_\alpha\}_{\alpha \in i} \subset A$, $i = \{1, \ldots, N_r\}$, and a partition $\{\mathcal{I}_i\}_{i=1}^{\mathfrak{D}_n}$ of i so that*

$$A \subset \bigcup_{\alpha \in i} B(x_\alpha, r), \qquad B(x_{\alpha_1}, 3r) \cap B(x_{\alpha_2}, 3r) = \emptyset, \qquad \alpha_1, \alpha_2 \in \mathcal{I}_i, \qquad \alpha_1 \neq \alpha_2,$$

$$\{x_\alpha\}_{\alpha \in i} \text{ is an } \frac{r}{2} \text{ maximal separated set in } A.$$

Proof Let $\{x_\alpha\}_{\alpha \in i}$ be a maximal $\frac{r}{2}$ separated set in A. Fix $\alpha_0 \in i$ and suppose that $B(x_{\alpha_0}, 3r) \cap B(x_\alpha, 3r) \neq \emptyset$ for all $\alpha \in \mathcal{K}_{\alpha_0} \subset i$. Then for all $\alpha \in \mathcal{K}_{\alpha_0}$, $B(x_\alpha, \frac{r}{2}) \subset B(x_{\alpha_0}, 8r)$. In particular,

$$\sum_{\alpha \in \mathcal{K}_{\alpha_0}} \operatorname{vol}(B(x_\alpha, \tfrac{r}{2})) \le \operatorname{vol}(B(x_{\alpha_0}, 8r)).$$

Now, there exist $R_0 > 0$ depending on n and a lower bound on the sectional curvature of $\tilde{M}$, and $\mathfrak{D}_n > 0$ depending only on n, so that for all $0 < r < R_0$,

$$\operatorname{vol}(B(x_{\alpha_0}, 8r)) \le \operatorname{vol}(B(x_\alpha, 14r)) \le \mathfrak{D}_n \operatorname{vol}(B(x_\alpha, \tfrac{r}{2})). \tag{6.15}$$

Hence, it follows from (6.15) that

$$\sum_{\alpha \in \mathcal{K}_{\alpha_0}} \mathrm{vol}(B(x_\alpha, \tfrac{r}{2})) \leq \mathrm{vol}(B(\rho_{\alpha_0}, 8r)) \leq \frac{\mathfrak{D}_n}{|\mathcal{K}_{\alpha_0}|} \sum_{\alpha \in \mathcal{K}_{\alpha_0}} \mathrm{vol}(B(x_\alpha, \tfrac{r}{2})).$$

In particular, $|\mathcal{K}_{\alpha_0}| \leq \mathfrak{D}_n$.

At this point we have proved that each of the balls $B(x_\alpha, 3r)$ intersects at most $\mathfrak{D}_n - 1$ other balls. We now construct the sets $\mathcal{I}_1, \ldots, \mathcal{I}_{\mathfrak{D}_n}$ using a greedy algorithm. We will say that the index α_1 *intersects* the index α_2 if

$$B(x_{\alpha_1}, 3r) \cap B(x_{\alpha_2}, 3r) \neq \emptyset.$$

We place the index $1 \in \mathcal{I}_1$. Then suppose we have placed the indices $\{1, \ldots, \alpha\}$ in $\mathcal{I}_1, \ldots, \mathcal{I}_{\mathfrak{D}_n}$ so each of the $\mathcal{I}_i$'s consists of disjoint indices. Then, since $\alpha + 1$ intersects at most $\mathfrak{D}_n - 1$ indices, it is disjoint from $\mathcal{I}_i$ for some i. We add the index α to $\mathcal{I}_i$. By induction we obtain the partition $\mathcal{I}_1, \ldots, \mathcal{I}_{\mathfrak{D}_n}$.

Now, suppose that there exists $x \in A$ so that $x \notin \bigcup_{\alpha \in i} B(x_\alpha, r)$. Then, $\min_{\alpha \in \mathcal{I}} d(x, x_\alpha) \geq r$, a contradiction of the $r/2$ maximality of x_α. $\square$

Our next lemma shows that there is $\mathfrak{D}_n > 0$ depending only on n such that one can construct a $(\mathfrak{D}_n, \tau, r)$-good cover of $\mathcal{H}_0$.

Lemma 6.2.6 (Good covers exist) *There exist $\mathfrak{D}_n > 0$ depending only on n, $R_0 > 0$ depending on n and a bound for the curvature of $\mathcal{S}$, such that for $0 < r_1 < R_0$, $0 < r_0 \leq \tfrac{r_1}{2}$, $A \subset \mathcal{H}_0$, there exist points $\{\rho_j\}_{j=1}^{N_{r_1}} \subset A$ and a partition $\{\mathcal{J}_i\}_{i=1}^{\mathfrak{D}_n}$ of $\{1, \ldots, N_{r_1}\}$ so that for all $0 < \tau < \frac{\tau_{injH}}{2}$*

- $\Lambda_A^\tau(r_0) \subset \bigcup_{j=1}^{N_{r_1}} \Lambda_{\rho_j}^\tau(r_1)$,
- $\Lambda_{\rho_j}^\tau(3r_1) \cap \Lambda_{\rho_\ell}^\tau(3r_1) = \emptyset, \qquad j, \ell \in \mathcal{J}_i, \quad j \neq \ell.$

Proof We first apply Lemma 6.2.5 to $\tilde{M} = \mathcal{H}_0$ to obtain $R_0 > 0$ depending only on n and a bound for the curvature of $\mathcal{S}$ so that for $0 < r_1 < R_0$, there exist $\{\rho_j\}_{j=1}^{N_{r_1}} \subset \mathcal{H}_0$ and a partition $\{\mathcal{J}_i\}_{i=1}^{\mathfrak{D}_n}$ of $\{1, \ldots, N_{r_1}\}$ such that

$$A \subset \bigcup_{j=1}^{N_{r_1}} B(\rho_j, r_1), \qquad B(\rho_j, 3r_1) \cap B(\rho_\ell, 3r_1) = \emptyset, \qquad j, \ell \in \mathcal{J}_i, \quad j \neq \ell,$$

$$\{\rho_j\}_{j=1}^{N_{r_1}} \text{ is an } \frac{r_1}{2} \text{ maximal separated set in } A.$$

Now, suppose that $j, \ell \in \mathcal{J}_i$ and

$$\Lambda^\tau_{\rho_\ell}(3r_1) \cap \Lambda^\tau_{\rho_j}(3r_1) \neq \emptyset.$$

Then, there exist $q_\ell \in B(\rho_\ell, 3r_1) \cap \mathcal{H}_\varepsilon$, $q_j \in B(\rho_j, 3r_1) \cap \mathcal{H}_\varepsilon$ and $t_\ell, t_j \in [-\tau, \tau]$ so that $\varphi_{t_\ell - t_j}(q_\ell) = q_j$. In particular, for $\tau < \tau_{\text{inj}H}/2$, this implies that $q_\ell = q_j$, $t_\ell = t_j$ and hence $B(\rho_\ell, 3r_1) \cap B(\rho_j, 3r_1) \neq \emptyset$ a contradiction.

Now, suppose $r_0 \leq r_1$ and that there exists $\rho \in \Lambda^\tau_A(r_0)$ so that $\rho \notin \bigcup_{j=1,\dots,N_{r_1}} \Lambda^\tau_{\rho_j}(r_1)$. Then, there are $|t| < \tau + r_0$ and $q \in \mathcal{S}$ so that

$$\rho = \varphi_t(q), \qquad d(q, A) < r_0, \qquad \min_{j=1,\dots,N_{r_1}} d(q, \rho_j) \geq r_1.$$

In particular, there exists $\tilde{\rho} \in A$ with $d(q, \tilde{\rho}) < r_0$ such that for all $j = 1, \dots, N_{r_1}$,

$$d(\tilde{\rho}, \rho_j) \geq d(q, \rho_j) - d(q, \tilde{\rho}) > r_1 - r_0.$$

This contradicts the maximality of $\{\rho_j\}_{j=1}^{N_{r_1}}$ if $r_0 \leq r_1/2$. $\qquad\qquad\square$

We proceed to build a δ-good partition of unity associated to the cover we constructed in Lemma 6.2.6. The key feature in this partition is that it is invariant under the geodesic flow. Indeed, the partition is built so that its quantization commutes with the operator $P = -h^2 \Delta - I$ in a neighborhood of $\mathcal{H}_0$.

Lemma 6.2.7 (Good partitions exist) *There exist $\tau_1 = \tau_1(\tau_{\text{inj}H}) > 0$ and $\varepsilon_1 = \varepsilon_1(\tau_1) > 0$, and given $0 < \delta < \frac{1}{2}$, $0 < \varepsilon \leq \varepsilon_1$, there exists $h_1 > 0$, so that for any $0 < \tau \leq \tau_1$, and $R(h) \geq h^\delta$, the following holds.*

There exist $C_1 > 0$ so that for all $0 < h \leq h_1$ and every $(\tau, R(h))$-cover of A there exists a partition of unity $\chi_j \in S_\delta \cap C_c^\infty(T^*M; [-C_1 h^{1-2\delta}, 1 + C_1 h^{1-2\delta}])$ on $\Lambda^\tau_A(\frac{1}{2} R(h))$ for which

$$\text{supp}\, \chi_j \subset \Lambda^{\tau+\varepsilon}_{\rho_j}(R(h)), \qquad \text{MS}_h([P, Op_h(\chi_j)]) \cap \Lambda^\tau_{\mathcal{H}_0}(\varepsilon) = \emptyset,$$

$$\sum_j \chi_j \equiv 1 \text{ on } \Lambda^\tau_A(\tfrac{1}{2} R(h)),$$

and $\{\chi_j\}_j$ is bounded in S_δ, and $[P, Op_h(\chi_j)]$ is bounded in Ψ_δ.

Remark 6.2.8 (*Decomposition into beams*) Since $\sum_j \chi_j \equiv 1$ near the τ flow-out of A by the geodesic flow, we will use Lemma 6.2.7 to decompose u microlocally near this flow-out as

$$u = \sum_j Op_h(\chi_j)u,$$

i.e. as a sum of geodesic beams.

Proof Let $\mathcal{S}$ as above, $\tau_1 < \frac{1}{2}\tau_{\mathrm{inj}H}$ and fix $0 < \tau \leq \tau_1$. Then let $\varepsilon_1 > 0$ be so small that $\Lambda_{\mathcal{H}_0}^{\tau_1}(\varepsilon_1) \subset \Lambda_{\mathcal{S}}^{2\tau_1}(0)$, fix $0 < \varepsilon < \varepsilon_1$ and let h_1 be so small that $h^\delta \leq \varepsilon$ for all $0 < h \leq h_1$. For each $j \in \{1, \ldots, N_h\}$ let $\mathcal{H}_j = \mathcal{S} \cap \Lambda_{\rho_j}^{\tau}(R(h))$. Let $\{\psi_j\} \subset C_c^\infty(\mathcal{S}; [0, 1]) \cap S_\delta$ be a partition of unity on $\mathcal{S} \cap \Lambda_A^{\tau}(\frac{1}{2}R(h))$ subordinate to $\{\mathcal{H}_j\}_{j=1}^{N_h}$ that is uniformly bounded in S_δ. Then, define $a_{j,0} \in S_\delta$ on $\Lambda_A^{\tau}(\varepsilon)$ by solving

$$a_{j,0}|_{\mathcal{L}_h} = \psi_j, \qquad H_p a_{j,0} = 0 \quad \text{on} \quad \Lambda_A^{\tau}(\varepsilon).$$

Clearly, $a_{j,0}$ defined in this way is a partition of unity for $\Lambda_A^{\tau}(\frac{1}{2}R(h))$. Furthermore, we can extend $a_{j,0}$ to T^*M as an element of S_δ so that

$$\operatorname{supp} a_{j,0} \subset \bigcup_{|t| \leq \tau + \varepsilon + R(h)} \varphi_t(\mathcal{H}_j) \subset \Lambda_{\rho_j}^{\tau+\varepsilon}(R(h)), \qquad 0 \leq a_{j,0} \leq 1$$

Note also that since $P \in \Psi^2(M)$ and $H_p a_{j,0} = 0$, for $b \in S_\delta$ with $\operatorname{supp} b \subset \Lambda_{\mathcal{H}_0}^{\tau}(\varepsilon)$,

$$Op_h(b)[P, Op_h(a_{j,0})] \in h^{2-2\delta}\Psi_\delta(M).$$

We define $a_{j,k}$ by induction. Suppose we have $a_{j,\ell}$, $\ell = 0, \ldots, k-1$, so that if we set $\chi_{j,k-1} := \sum_{\ell=0}^{k-1} h^{\ell(1-2\delta)} a_{j,\ell}$, then

(A) $\displaystyle\sum_{j=1}^{N_h} \chi_{j,k-1} \equiv 1 \text{ on } \Lambda_{\mathcal{H}_0}^{\tau}(\tfrac{1}{2}R(h))$,

(B) $e_{j,k} := \sigma\Big(h^{-1-k(1-2\delta)}[P, Op_h(\chi_{j,k-1})]\Big) \in S_\delta \quad \text{on } \Lambda_{\mathcal{H}_0}^{\tau}(\varepsilon)$.

Then, for every $k \geq 1$ define $a_{j,k} \in S_\delta$ by

$$a_{j,k}|_{\mathcal{L}_h} = 0, \qquad H_p a_{j,k} = -i e_{j,k} \quad \text{on} \quad \Lambda_{\mathcal{H}_0}^{\tau}(\varepsilon). \tag{6.16}$$

Next extend $a_{j,k}$ to T^*M as an element of S_δ so that

$$\operatorname{supp} a_{j,k} \subset \bigcup_{|t| \leq \tau + \varepsilon + R(h)} \varphi_t(\mathcal{H}_j) \subset \Lambda_{\rho_j}^{\tau+\varepsilon}(R(h)).$$

Now, since $\sum_{j=1}^{N_h} \chi_{j,k-1} \equiv 1$ on $\Lambda_{\mathcal{H}_0}^{\tau}(\frac{1}{2}R(h))$, by (B) we see that for $\rho \in \Lambda_{\mathcal{H}_0}^{\tau}(\frac{1}{2}R(h))$,

$$\sum_{j=1}^{N_h} e_{j,k}(\rho) = \sigma\Big(h^{-1-k(1-2\delta)}\Big[P, Op_h\Big(\sum_{j=1}^{N_h} \chi_{j,k-1}\Big)\Big]\Big)(\rho) = 0.$$

In particular, (6.16) gives that $\sum_{j=1}^{N_h} a_{j,k} = 0$ on $\Lambda_{\mathcal{H}_0}^{\tau}(\frac{1}{2}R(h))$. Therefore, since $\chi_{j,k} = \chi_{j,k-1} + h^{k(1-2\delta)} a_{j,k}$, we conclude that

$$\sum_{j=1}^{N_h} \chi_{j,k} = 1 \quad \text{on} \ \ \Lambda^\tau_{\mathcal{H}_0}(\tfrac{1}{2}R(h)),$$

and hence (A) is satisfied for $a_{j,\ell}$ with $\ell = 0, \dots, k$. To show that (B) is also satisfied, let $b \in S_\delta$ with $\operatorname{supp} b \subset \Lambda^\tau_{\mathcal{H}_0}(\varepsilon)$. By assumption, we have

$$Op_h(b)[P, Op_h(\chi_{j,k-1})] \in h^{1+k(1-2\delta)}\Psi_\delta(M).$$

Also, using once again that $P \in \Psi^m(M)$ and that $H_p a_{j,k} = -i e_{j,k}$

$$Op_h(b)[P, Op_h(a_{j,k})] \in h\Psi_\delta(M) + h^{2-2\delta}\Psi_\delta(M).$$

Hence,

$$Op_h(b)[P, Op_h(\chi_{j,k})] = Op_h(b)\big[P, Op_h(\chi_{j,k-1} + h^{k(1-2\delta)}a_{j,k})\big] \in h^{1+k(1-2\delta)}\Psi_\delta(M),$$

and so, on $\Lambda^\tau_{\mathcal{H}_0}(\varepsilon)$,

$$\begin{aligned}
\sigma(h^{-1-k(1-2\delta)}Op_h(b)[P, Op_h(\chi_{j,k})]) &= \\
&= \sigma\Big(h^{-1-k(1-2\delta)}Op_h(b)\big([P, Op_h(\chi_{j,k-1})] + h^{k(1-2\delta)}[P, Op_h(a_{j,k})]\big)\Big) \\
&= b(e_{j,k} - e_{j,k}) = 0.
\end{aligned}$$

In particular,
$$Op_h(b)[P, Op_h(\chi_{j,k})] \in h^{1+(k+1)(1-2\delta)}\Psi_\delta(M), \tag{6.17}$$

and $e_{j,k+1} \in S_\delta$ on $\Lambda^\tau_{\mathcal{H}_0}(\varepsilon)$ as claimed.

Finally, let
$$\chi_j \sim \sum_{\ell=0}^{\infty} h^{\ell(1-2\delta)}a_{j,\ell}.$$

Then, using (6.17),
$$\mathrm{MS_h}([P, Op_h(\chi_j)]) \cap \Lambda^\tau_{\mathcal{H}_0}(\varepsilon) = \emptyset.$$

Now, note that by construction $\{\chi_j\}$ remains a partition of unity modulo $O(h^\infty)$ and by adding an h^∞ correction to teach term, we construct $\{\chi_j\}$ so that it forms a partition of unity. We also have by construction that $\chi_j \in C_c^\infty(T^*M; [-C_1 h^{1-2\delta}, 1 + C_1 h^{1-2\delta}])$ for some C_1 depending only on (M, p) and curvature bounds for $\mathcal{S}$. $\qquad\square$

Applications of the Geodesic Beam Decomposition 7

This chapter illustrates how to apply the geodesic beam techniques developed in Chap. 6 to find effective pointwise bounds for Laplace eigenfunctions (Sect. 7.1), control averages of eigenfunctions over submanifolds (Sect. 7.2), bound L^p norms of eigenfunctions for $p > p_c$ (Sect. 7.3) and find improvements on the remainder for the Weyl Law (see Sect. 7.4).

7.1 Applications to Sup-Norms

7.1.1 The L^∞ Estimate in Terms of Analytic Data

Our first application of the geodesic beam techniques is the key estimate we use to study L^∞ estimates. Its analog for averages will be given in the next section. This estimate controls the L^∞ norm of an eigenfunction (or quasimode) for the Laplacian in terms of the L^2 mass of the geodesic beams running over that ball as well as an error microlocalized away from S^*M. This estimate is useful in applications because, via an application of Egorov's theorem, it can be used to control the values of u near x using only dynamical information about the geodesic flow (see Theorem 7.1.5). In addition, this estimate describes the structure that an eigenfunction must have near a point x in order saturate the standard L^∞ bound at x. Indeed, it is not hard to check from (7.1) that for the L^∞ norm bound to be saturated the distribution of mass (at scale h^δ) among all geodesics passing through x is nearly uniform i.e. u looks similar to a Zonal harmonic (see Remark 7.1.2).

Throughout this section, we use the fact that for every $x \in M$, there is a hypersurface $\mathcal{S}_x \subset S^*M$ transverse to $H_{|\xi|_g}$ such that $S_x^*M \subset \mathcal{S}_x$ and the curvature of $\mathcal{S}_x$ is uniformly bounded. For a given x, we use the surface $\mathcal{S}_x$ to define geodesic tubes.

Y. Canzani and J. Galkowski, *Geodesic Beams in Eigenfunction Analysis*, Synthesis Lectures on Mathematics & Statistics, https://doi.org/10.1007/978-3-031-31586-2_7

Theorem 7.1.1 *Let $x \in M$. There exist $\tau_0 = \tau_0(M, g) > 0$, $R_0 = R_0(M, g) > 0$, $C_n > 0$ depending only on n, so that the following holds.*

Let $0 < \tau \leq \tau_0$, $0 \leq \delta < \frac{1}{2}$, and $R(h) := \min(h^\delta, R_0)$. Let $\{\chi_j\}_{j \in \mathcal{J}}$ be a δ-partition for $S_x^* M$ associated to a $(\tau, R(h))$-cover $\{\Lambda_{\rho_j}^\tau(R(h))\}_{j \in \mathcal{J}}$. Let $N > 0$ and $s > \frac{n}{2} - 2$, and $e \in S_\delta$ with

$$\operatorname{supp}(1 - e) \cap \{\rho \in T^*M \; : \; d(\rho, S^*M) > \tfrac{1}{8} R(h)\} = \emptyset.$$

Then, there are $h_0 = h_0(M, g, \{\chi_j\}, \delta) > 0$, $C_s > 0$ and $C_N > 0$ with the property that for any $0 < h < h_0$ and $u \in \mathcal{D}'(M)$

$$\|u\|_{L^\infty(B(x, \frac{1}{16} R(h)))}$$
$$\leq C_n \tau^{-\frac{1}{2}} h^{\frac{1-n}{2}} R(h)^{\frac{n-1}{2}} \sum_{j \in \mathcal{J}} \left(\|Op_h(\chi_j) u\|_{L^2} + h^{-1} \|Op_h(\chi_j)(-h^2 \Delta_g - 1) u\|_{L^2} \right)$$
$$+ C_s h^{-1} h^{-\frac{n-1}{2}+1} R(h)^{-1} \|Op_h(e)(-h^2 \Delta_g - 1) u\|_{H_h^s M)} + C_N h^N \|u\|_{L^2}.$$

$$(7.1)$$

Moreover, the constants h_0 and C_N are uniform for χ_j in bounded subsets of S_δ.

Remark 7.1.2 One can conclude from Theorem 7.1.1 that, in order to have maximal sup-norm growth at a point, an eigenfunction, u_h, must have a component with L^2 norm bounded from below that is distributed in the same way as the canonical Zonal harmonic example on the sphere -up to scale h^δ for all $\delta < \frac{1}{2}$-(see Fig. 2.1). Indeed, if one restricts attention to (τ, r) covers of $S_x^* M$ without too many overlaps (see Definition 6.2.4) it follows from Theorem 7.1.1 that there exists $C_n > 0$, so that for all $\varepsilon > 0$, if

$$\#\left\{ j \; : \; \varepsilon^2 R(h)^{n-1} \leq \|Op_h(\chi_j) u_h\|_{L^2(M)}^2 \leq \frac{R(h)^{n-1}}{\varepsilon^2} \right\} \leq \varepsilon^2 N_h,$$

then $\|u_h\|_{L^\infty(B(x, h^\delta))} \leq \varepsilon C_n \tau^{-\frac{1}{2}} h^{\frac{1-n}{2}}$.

To understand Theorem 7.1.1 more heuristically, one should think of $\|Op_h(\chi_j) u_h\|_{L^2(M)}$ as measuring the L^2 mass of u_h on the tube of radius $R(h)$ around a geodesic that runs through the point x. Since $\operatorname{vol}_{S_x^* M}(\operatorname{supp} \chi_j) \asymp R(h)^{n-1}$, an individual term in the sum in Theorem 7.1.1 is then

$$R(h)^{\frac{n-1}{2}} \|Op_h(\chi_j) u_h\|_{L^2(M)} \asymp \left(\frac{\|Op_h(\chi_j) u_h\|_{L^2(M)}^2}{\operatorname{vol}_{S_x^* M}(\operatorname{supp} \chi_j)} \right)^{\frac{1}{2}} \operatorname{vol}_{S_x^* M}(\operatorname{supp} \chi_j),$$

where $\operatorname{vol}_{S_x^* M}$ is the volume measure on $S_x^* M$ induced by the Sasaki metric on T^*M. In particular, the sum on the right of the estimate in Theorem 7.1.1 can be interpreted as $\int_{S_x^* M} |\frac{d\mu}{d \operatorname{vol}_{S_x^* M}}|^{\frac{1}{2}} d \operatorname{vol}_{S_x^* M}$, where μ is the measure giving the distribution of the mass squared of u_h on $S_x^* M$. This statement can be made precise by using defect measures (see [CG19b, Theorem 6] or [Gal19]), but the results using defect measures can only be used to obtain $o(1)$ improvements on eigenfunction bounds.

Proof of Theorem 7.1.1. Let $\{\mathcal{T}_i\}_{i\in\mathcal{I}}$ be a good cover of $S_x^* M$ by tubes of length $\tau/2$ and radius $R(h)/2$, i.e. there are $\{q_i\}_{i\in\mathbf{i}}$ such that $\mathcal{T}_i = \Lambda_{q_i}^{\tau/2}(R(h)/2)$. Let $\{X_{\mathcal{T}_i}\}_{i\in\mathcal{I}}$ a good partition near $S_x^* M$ subordinate to this cover. Next, let $\chi \in C_c^\infty((-2,2))$ with $\chi \equiv 1$ on $[-1,1]$, and define the 'local' and 'far' localizers as

$$\chi_L(y) := \chi(16R(h)^{-1}d(y,x))), \qquad \chi_F(y) := 1 - \chi_L(y).$$

In addition, let

$$\Phi_L := \chi_L\Big(1 - \sum_{i\in\mathcal{I}} X_{\mathcal{T}_i}\Big), \qquad \Phi_F := \chi_F\Big(1 - \sum_{i\in\mathcal{I}} X_{\mathcal{T}_i}\Big).$$

Then, we claim that

$$\mathrm{MS}_h(\Phi_L) \subset \{\rho \in T^*M \ : \ d(\rho, S^*M) \geq \tfrac{1}{8}R(h)\}. \tag{7.2}$$

To see the inclusion, observe that

$$\mathrm{MS}_h\Big(\chi_L(I - \sum_{i\in\mathcal{I}} X_{\mathcal{T}_i})\Big) \subset T^*_{B(x,\frac{1}{8}R(h))}M \cap \Big(T^*M \setminus \Lambda_{S_x^*M}^{\tau/2}(\tfrac{1}{2}R(h))\Big) = \emptyset.$$

Suppose that $\rho \in T^*_{B(x,\frac{1}{8}R(h))}M$ with $d(\rho, S^*M) \leq \tfrac{1}{8}R(h)$. Then, $d(\rho, S_x^*M) \leq \tfrac{1}{4}R(h)$. In particular, since for h small enough $\Lambda_{S_x^*M}^{\tau/2}(\tfrac{1}{2}R(h)) \supset B(S_x^*M, \tfrac{1}{4}R(h))$, we have $\rho \in \Lambda_{S_x^*M}^{\tau/2}(\tfrac{1}{2}R(h))$ and hence the claim in (7.2) follows.

We then decompose

$$u := \sum_{i\in\mathcal{I}} X_{\mathcal{T}_i}u + \Phi_L u + \Phi_F u,$$

so that

$$\|u\|_{L^\infty(B(x,\frac{1}{16}R(h)))} \leq \sum_{i\in\mathcal{I}} \|X_{\mathcal{T}_i}u\|_{L^\infty(B(x,\frac{1}{16}R(h)))} + \|\Phi_L u\|_{L^\infty(B(x,\frac{1}{16}R(h)))} \\ + \|\Phi_F u\|_{L^\infty(B(x,\frac{1}{16}R(h)))}, \tag{7.3}$$

Observe that by construction,

$$\|\Phi_F u\|_{L^\infty(B(x,\frac{1}{16}R(h)))} = 0, \tag{7.4}$$

and hence we only need to estimate the remaining two terms.

Next, by Lemmas 6.1.1 and 6.2.2, we have for $h^\delta < R(h) < R_0$ and $0 < \tau < \tau_0$,

$$\|X_{\mathcal{T}_i}u\|_{L^\infty(B(x,\frac{1}{16}R(h)))}$$

$$\leq \frac{C_n}{\tau^{1/2}}h^{\frac{1-n}{2}}R(h)^{\frac{n-1}{2}}\Big(\|X_{\mathcal{T}_i}u\|_{L^2} + h^{-1}\|X_{\mathcal{T}_i}(-h^2\Delta_g - 1)u\|_{L^2} + C_N h^N\|u\|_{L^2}\Big). \tag{7.5}$$

Finally, we use Lemma 3.3.3 to control $\Phi_L u$. Indeed, for any $e \in S^0_\delta$ with $e \equiv 1$ on

$$\{d(\rho, S^*M) : \rho > \tfrac{1}{4} R(h)\},$$

Lemma 3.3.3 implies that for any $s \in \mathbb{R}$,

$$\|\Phi_L u\|_{H^s_h} \leq C_s R(h)^{-1} \|Op_h(e)(-h^2 \Delta - 1)u\|_{H^{s-2}_h} + C_N h^N \|u\|_{L^2},$$

and hence that for $s > \tfrac{n}{2}$,

$$\|\Phi_L u\|_{L^\infty} \leq C_s R(h)^{-1} h^{-\frac{n}{2}} \|Op_h(e)(-h^2 \Delta - 1)u\|_{H^{s-2}_h} + C_N h^N \|u\|_{L^2}, \tag{7.6}$$

All together, (7.3), (7.4), (7.5), and (7.6) yield

$$\|u\|_{L^\infty(B(x, \frac{1}{16} R(h)))} \leq \frac{C_n}{\tau^{1/2}} h^{\frac{1-n}{2}} R(h)^{\frac{n-1}{2}} \sum_{i \in \mathbf{i}} (\|X_{T_i} u\|_{L^2} + h^{-1} \|X_{T_i}(-h^2 \Delta_g - 1)u\|_{L^2})$$

$$+ C_s R(h)^{-1} h^{-\frac{n}{2}} \|Op_h(e)(-h^2 \Delta - 1)u\|_{H^{s-2}_h} + C_N h^N \|u\|_{L^2}. \tag{7.7}$$

Next, observe that by the definition of a δ-partition (see Definition 6.2.3), for any $i \in \mathcal{I}$, $\sum_j \chi_j \geq 1$ on T_i and hence, since $\{X_{T_i}\}_{i \in \mathcal{I}}$ is bounded in Ψ_δ, there is $\{E_i\}_{i \in \mathbf{i}}$ bounded in Ψ_δ with $\sigma(E_i) \leq 1$ on T_i such that for each $i \in \mathcal{I}$

$$X_{T_i} = \sum_{j \in \mathcal{J}} X_{T_i} E_i Op_h(\chi_j) + O(h^\infty)_{\Psi^{-\infty}},$$

and hence, defining

$$\mathcal{A}_i = \{j \in \mathcal{J} : \Lambda^\tau_{\rho_j}(R(h)) \cap \Lambda^{\tau/2}_{q_i}(\tfrac{1}{2} R(h)) \neq \varnothing\},$$

we have

$$X_{T_i} = \sum_{j \in \mathcal{A}_i} X_{T_i} E_i Op_h(\chi_j) + O(h^\infty)_{\Psi^{-\infty}}.$$

Therefore,

$$\sum_{i \in \mathcal{I}} \|X_{T_i} v\|_{L^2} \leq \sum_{i \in \mathcal{I}} \sum_{j \in \mathcal{A}_i} \|X_{T_i} E_i Op_h(\chi_j)v\|_{L^2} + C_N h^N \|v\|_{L^2}$$

$$= \sum_{j \in \mathcal{J}} \sum_{i \in \mathcal{B}_j} \|X_{T_i} E_i Op_h(\chi_j)v\|_{L^2} + C_N h^N \|v\|_{L^2} \tag{7.8}$$

$$= 2 \sup_{j \in \mathcal{J}} |\mathcal{B}_j| \sum_{j \in \mathcal{J}} \|Op_h(\chi_j)v\|_{L^2} + C_N h^N \|v\|_{L^2},$$

where

$$\mathcal{B}_j := \{i \in \mathcal{I} \, : \, \Lambda_{q_i}^{\tau/2}(\tfrac{1}{2}R(h)) \cap \Lambda_{\rho_j}^{\tau}(R(h)) \neq \emptyset\}.$$

To complete the proof, we claim that there is $C_n > 0$ such that

$$\sup_{j \in \mathcal{J}} |\mathcal{B}_j| \leq C_n. \tag{7.9}$$

To see (7.9), observe that for $i \in \mathcal{B}_j$,

$$\Lambda_{\rho_j}^{\tau}(R(h)) \subset \Lambda_{q_i}^{\tau}(3R(h)).$$

Therefore,

$$\mathcal{B}_j \subset \{i_1 \in \mathcal{I} \, : \, \Lambda_{q_i}^{\tau}(3R(h)) \cap \Lambda_{q_{i_1}}^{\tau}(3R(h))\},$$

and, since

$$\#\{i_1 \in \mathcal{I} \, : \, \Lambda_{q_i}^{\tau}(3R(h)) \cap \Lambda_{q_{i_1}}^{\tau}(3R(h))\} \leq C_n,$$

the claim (7.9) follows.

Using (7.8) in (7.8) and applying this to the first two terms in (7.7), we obtain

$$\|u\|_{L^\infty(B(x, \frac{1}{16}R(h)))}$$
$$\leq \frac{C_n}{\tau^{1/2}} h^{\frac{1-n}{2}} R(h)^{\frac{n-1}{2}} \sum_{j \in \mathcal{J}} (\|Op_h(\chi_j)u\|_{L^2} + h^{-1}\|Op_h(\chi_j)(-h^2\Delta_g - 1)u\|_{L^2})$$
$$+ C_s R(h)^{-1} h^{-\frac{n}{2}}\|Op_h(e)(-h^2\Delta_g - 1)u\|_{H_h^{s-2}}$$
$$+ C_N h^N \|u\|_{L^2},$$

which is the desired estimate. $\qquad\square$

7.1.2 Error Term as a Function of the Laplacian

Before proceeding to estimates on L^∞ norms in terms of dynamical data, we prove a lemma which shows that the error term

$$\|Op_h(e)(-h^2\Delta_g - 1)u\|_{H_h^{s-2}}$$

appearing in Theorem 7.1.1 may be replaced by an error of the form

$$\left\|\left(1 - \psi\left(\tfrac{-h^2\Delta_g - 1}{h^\delta}\right)\right)(-h^2\Delta_g - 1)u\right\|_{H_h^{s-2}} + C_N h^N \|u\|_{H_h^{-N}},$$

where $\psi \in C_c^\infty(\mathbb{R})$ is identically 1 in a neighborhood of 0. When working with spectral projectors, it is more convenient to have this type of error term since one understands well norms of the operator

$$\Big(1 - \psi\big(\tfrac{-h^2\Delta_g - 1}{h^\delta}\big)\Big)f(-h^2\Delta_g)$$

for any $f \in L^\infty(\mathbb{R})$.

Lemma 7.1.3 *Let $e \in \Psi_\delta(M)$, $\psi \in C_c^\infty(\mathbb{R})$, and define*

$$\psi_{h,\delta}(x,\xi) := \psi\Big(h^{-\delta}(|\xi|^2_{g(x)} - 1)\Big).$$

Suppose that there is $c > 0$ such that for $0 < h < 1$,

$$d\big(\operatorname{supp} e, \operatorname{supp}\psi_{h,\delta}\big) > ch^\delta. \tag{7.10}$$

Then for all $N > 0$ there is $C_N > 0$ such that for all $0 < h < 1$,

$$\|Op_h(e)\psi\big(\tfrac{-h^2\Delta_g - 1}{h^\delta}\big)\|_{H_h^{-N} \to H_h^N} \le C_N h^N.$$

Proof Let $0 < a < b$. Then, by Lemma 3.3.3, for all $E \in [a, b]$ and $c > 0$, and any bounded subset $\mathcal{W}$ of S_δ and $N > 0$, there is a bounded subset $\mathcal{V}$ of $h^{-\delta}\Psi_\delta$ and $C > 0$ such that for any $e \in \mathcal{W}$ satisfying

$$d\big(\operatorname{supp} e, \{|\xi|^2_g = \lambda\}\big) \ge ch^\delta,$$

there is $A_E \in \mathcal{V}$ such that

$$\|Op_h(e) - A_E(-h^2\Delta_g - \lambda)\|_{H_h^{-N} \to H_h^N} \le Ch^N.$$

Now, observe that

$$Op_h(e)\psi\Big(\tfrac{-h^2\Delta_g - 1}{h^\delta}\Big)v = \sum_j \psi\big(\tfrac{E_j(h) - 1}{h^\delta}\big)\langle v, u_{E_j}\rangle_{L^2} Op_h(e)u_{E_j},$$

where $\{u_{E_j}\}_{j=1}^\infty$ is an orthonormal basis of eigenfunctions for the Laplacian satisfying

$$(-h^2\Delta_g - E_j(h))u_{E_j} = 0.$$

By (7.10), there is $c > 0$ such that for all j satisfying

$$\psi\big(\tfrac{E_j(h) - 1}{h^\delta}\big) \ne 0, \qquad \text{we have} \qquad d(\operatorname{supp} e, \{|\xi|^2_g - E_j(h)\}) \ge ch^\delta. \tag{7.11}$$

Therefore, there is a bounded subset $\mathcal{V}$ of $h^{-\delta}\Psi_\delta$ and $C > 0$ such that for all j satisfying (7.11), there is $A_{E_j} \in \mathcal{V}$ such that

$$\|Op_h(e)u_{E_j}\|_{L^2} = \|Op_h(e)u_{E_j} - A_{E_j}(-h^2\Delta_g - E_j)u_{E_j}\|_{L^2} \le Ch^K\|u_{E_j}\|_{L^2}.$$

In particular,

$$\left\| Op_h(e)\psi\left(\tfrac{-h^2\Delta_g-1}{h^\delta}\right)v\right\|_{H_h^N} \leq \sum_{|E_j-1|\leq Ch^\delta} Ch^K |\langle v, u_{E_j}\rangle_{L^2}|$$

$$\leq \sum_{|E_j-1|\leq Ch^\delta} Ch^K \|u_{E_j}\|_{H_h^N}\|v\|_{H_h^{-N}}$$

$$\leq \#\{j\ :\ |E_j-1|\leq Ch^\delta\}Ch^K\|v\|_{H_h^{-N}}$$

$$\leq Ch^{K-n}\|v\|_{H_h^{-N}}.$$

Choosing $K = N + n$ then proves the lemma. $\qquad\qquad\qquad\qquad\qquad\qquad\square$

7.1.3   The L^∞ Estimate in Terms of Dynamical Data

We now present an estimate on the L^∞ norm of a quasimode near a point that requires only dynamical information about geodesics which pass near that point. In concrete situations, this is often the more useful estimate since it is, apriori, one of the only ways we have to control the geodesic beam masses $\|Op_h(\chi_j)u\|_{L^2}$ in (7.1).

Before we proceed to this theorem, we record a useful application of Egorov's theorem in a technical lemma. We will need the concept of non-self looping to state it. A set $A \subset T^*M$ is $[t, T]$ non-self looping if either

$$\bigcup_{s\in[t,T]} \varphi_s(A)\cap A = \emptyset \qquad \text{or} \qquad \bigcup_{-s\in[t,T]} \varphi_s(A)\cap A = \emptyset. \qquad (7.12)$$

Lemma 7.1.4 *Let $0 \leq \delta < \tfrac{1}{2}$, and $0 < \gamma \leq \tfrac{1}{2} - \delta$, and $0 < \varepsilon < 2\gamma$. Then there is $\varepsilon_1 > 0$ such that for all $0 < t < T \leq (2\gamma - \varepsilon)\log h^{-1}$ and all $a \in S_\delta^{\mathrm{comp}}(T^*M;[0,1])$ with*

$$\mathrm{supp}\,a \text{ is } [t, T] \text{ is non-self looping}, \qquad \mathrm{supp}\,a \subset \{1 - \varepsilon_1 < |\xi|_g < 1 + \varepsilon_1\},$$

for $u \in \mathcal{D}'(M)$ we have

$$|\langle Op_h(a)u, u\rangle| \leq C\left(\frac{t}{T} + O(h^{1-2\gamma})\right)\|u\|_{L^2}^2 + \frac{Tt}{h^2}\|(-h^2\Delta_g - 1)u\|_{L^2}^2.$$

Proof We will assume that the first condition in (7.12) is satisfied. The proof in the other case is identical. We first write

$$\langle Op_h(a)u, u\rangle = \frac{1}{T}\int_0^T \langle Op_h(a)u, u\rangle ds$$

$$= \frac{1}{T}\int_0^T \langle e^{-ish\Delta_g}Op_h(a)e^{ish\Delta_g}u, u\rangle ds$$

$$- \frac{i}{Th}\int_0^T\int_0^s hD_r\langle e^{-irh\Delta_g}Op_h(a)e^{irh\Delta_g}u, u\rangle dr ds$$

$$=: I + II.$$

For the first term on the right, we use Theorem 3.4.1 to see that for all $s \in [0, T]$, there is $a_s \in S_\gamma$ with $\sigma(Op_h(a_s)) = \sigma(Op_h(a)) \circ \varphi_s$ (with $\{a_s\}_{s\in[0,T]}$ uniformly bounded in S_γ) such that

$$I = \left\langle \left(Op_h\left(\int_0^T \frac{1}{T}a_s ds\right) + O(h^\infty)_{\Psi-\infty}\right)u, u\right\rangle.$$

Now, since $\{a_s\}_{s\in[0,T]} \subset S_\gamma^{\mathrm{comp}}$ is uniformly bounded, we have

$$\int_0^T \frac{1}{T}a_s ds \in S_\gamma^{\mathrm{comp}}.$$

Thus, we have

$$\sigma\left[Op_h\left(\int_0^T \frac{1}{T}a_s ds\right)\right] = \int_0^T \frac{1}{T}\sigma(a_r)dr = \int_0^T \frac{1}{T}\sigma(Op_h(a)) \circ \varphi_s ds.$$

Finally, using that $\operatorname{supp} a$ is $[t, T]$ non-self looping, we obtain

$$\left|\int_0^T \frac{1}{T}\sigma(Op_h(a)) \circ \varphi_r dr\right| \leq \frac{t}{T}\sup|\sigma(Op_h(a))|.$$

Using this to estimate I, we obtain

$$|I| \leq \left(\frac{t}{T}\sup|\sigma(Op_h(a))| + O(h^{1-2\gamma})\right)\|u\|_{L^2}^2.$$

In fact, one can do slightly better in the estimate of I by using the fact that $\operatorname{ess\,supp} a_s \subset \varphi_{-s}(\operatorname{ess\,supp} a)$. Using this, we could replace the remainder by $O(tT^{-1}h^{1-2\gamma})$.

Next, we estimate the term II. Observe that

$$hD_s\langle e^{-ish\Delta_g}Op_h(a)e^{ish\Delta_g}u, u\rangle$$
$$= -\langle e^{-ish\Delta_g}Op_h(a)e^{ish\Delta_g}(-h^2\Delta_g - 1)u, u\rangle \tag{7.13}$$
$$+ \langle e^{-ish\Delta_g}Op_h(a)e^{ish\Delta_g}u, (-h^2\Delta_g - 1)u\rangle.$$

Using (7.13), we obtain

$$II = \frac{i}{Th} \int_0^T \int_0^s h D_r \langle e^{-irh\Delta_g} Op_h(a) e^{irh\Delta_g} u, u \rangle \, dr \, ds$$

$$= \frac{i}{Th} \int_0^T (T-r)\big(\langle e^{-irh\Delta_g} Op_h(a) e^{irh\Delta_g} u, (-h^2\Delta_g - 1)u \rangle$$

$$- \langle e^{-irh\Delta_g} Op_h(a) e^{irh\Delta_g}(-h^2\Delta_g - 1)u, u \rangle \big) \, dr$$

Then, using Theorem 3.4.1 to write $e^{-irh\Delta_g} Op_h(a) e^{irh\Delta_g} = Op_h(a_r) + O(h^\infty)_{\Psi^{-\infty}}$, we obtain

$$\frac{1}{Th}\left| \int_0^T (T-r)\big(\langle e^{-irh\Delta_g} Op_h(a) e^{irh\Delta_g} u, (-h^2\Delta_g - 1)u \rangle \, dr \right|$$

$$\leq \frac{1}{Th}\left| \left\langle Op_h\left(\int_0^T (T-r)a_r \, dr \right) u, (-h^2\Delta_g - 1)u \right\rangle \right| + O(h^\infty)\|u\|_{L^2}^2$$

Similarly,

$$\frac{1}{Th}\left| \int_0^T (T-r)\big(\langle e^{-irh\Delta_g} Op_h(a) e^{irh\Delta_g}(-h^2\Delta_g - 1)u, u \rangle \, dr \right|$$

$$\leq \frac{1}{Th}\left| \left\langle Op_h\left(\int_0^T (T-r)a_r \, dr \right)(-h^2\Delta_g - 1)u, u \right\rangle \right| + O(h^\infty)\|u\|_{L^2}^2$$

Now, uniformly for $0 \leq r \leq T$, a_r, $\frac{T-r}{T}a_r \in S_\gamma^{\mathrm{comp}}$. In particular,

$$\int_0^T \frac{(T-r)}{T} a_r \, dr \in S_\gamma^{\mathrm{comp}}.$$

Thus, we have

$$\sigma\left[Op_h\left(\int_0^T \frac{(T-r)}{T} a_r \, dr \right) \right] = \int_0^T \frac{T-r}{T} \sigma(a_r) \, dr = \int_0^T \frac{T-r}{T} \sigma(Op_h(a)) \circ \varphi_r \, dr.$$

Using again that $\mathrm{supp}\, a$ is $[t, T]$ non-self looping, we obtain

$$\left| \int_0^T \frac{T-r}{T} \sigma(Op_h(a)) \circ \varphi_r \, dr \right| \leq t \sup_{T^*M} |\sigma(Op_h(a))|.$$

Applying this estimate above, we have

$$|II| \leq \left(\frac{2t}{h} + O(h^{-2\gamma}) \right) \|u\|_{L^2} \|(-h^2\Delta_g - 1)u\|_{L^2} \sup_{T^*M} |\sigma(Op_h(a))|$$

$$\leq \sup_{T^*M} |\sigma(Op_h(a))| \left[\left(\frac{t}{T} + O(h^{2-4\gamma}T^{-1}) \right) \|u\|_{L^2}^2 + \frac{Tt}{h^2} \|(-h^2\Delta_g - 1)u\|_{L^2}^2 \right]$$

We note that, as in the estimate for I, we could use the finer property that $\mathrm{ess\,supp}(a_s) \subset \varphi_{-s}\, \mathrm{ess\,supp}(a)$ to improve the remainder in the estimate for II. $\qquad\square$

We now record the key dynamical estimate for dealing with sup-norm bounds. In this estimate, we control the L^∞ norm at a point by the looping structure of the geodesics through that point. We refer the reader to Chap. 8 for a discussion of how this estimate is used in practice and what its dynamical content is (See Fig. 7.1 for a picture of the covers used in Theorem 7.1.5 in the case of the square flat torus.)

Theorem 7.1.5 *Let $x \in M$, $0 < \delta < \frac{1}{2}$. There exist positive constants $h_0 = h_0(M, g, \delta)$, $\tau_0 = \tau_0(M, g)$, $R_0 = R_0(M, g)$, and C_n depending only on n, so that for all $0 < \tau \le \tau_0$ and $0 < h < h_0$ the following holds.*

Let $R(h) := \min(h^\delta, R_0)$, and $\{\Lambda_{\rho_j}^\tau(R(h))\}_{j=1}^{N_h}$ be a $(\tau, R(h))$-cover for $S_x^* M$. Let $0 \le \gamma < \frac{1}{2} - \delta$ and suppose there exists a partition of $\{1, \ldots, N_h\}$ into $\mathcal{B}$ and $\{\mathcal{G}_\ell\}_{\ell \in \mathcal{L}}$ such that for every $\ell \in \mathcal{L}$ there exist $T_\ell = T_\ell(h) > 0$ and $t_\ell = t_\ell(h) > 0$ with $t_\ell(h) \le T_\ell(h) \le \gamma T_e(h)$ such that

$$\bigcup_{j \in \mathcal{G}_\ell} \Lambda_{\rho_j}^\tau(R(h)) \quad is \quad [t_\ell, T_\ell] \text{ non-self looping.}$$

Then, for all $N > 0$ there exists $C_N = C_N(M, g, N, \tau, \delta) > 0$ so that for $u \in \mathcal{D}'(M)$

$$\|u\|_{L^\infty(B(x, \frac{1}{16} h^\delta))}$$

$$\le C_n \tau^{-\frac{1}{2}} h^{\frac{1-n}{2}} R(h)^{\frac{n-1}{2}} \left(|\mathcal{B}|^{\frac{1}{2}} \|u\|_{L^2} + \sum_{\ell \in \mathcal{L}} \frac{|\mathcal{G}_\ell|^{\frac{1}{2}} t_\ell^{\frac{1}{2}}}{T_\ell^{\frac{1}{2}}} \|u\|_{L^2} \right.$$

$$\left. + (|\mathcal{G}_\ell| t_\ell T_\ell)^{\frac{1}{2}} h^{-1} \|(-h^2 \Delta_g - 1)u\|_{L^2} \right)$$

$$+ C_n \tau^{-\frac{1}{2}} h^{-\frac{1-n}{2}} \|(-h^2 \Delta_g - 1)u\|_{L^2} + C_s h^{-\frac{n}{2}} R(h)^{-1} \|(-h^2 \Delta - 1)u\|_{H_h^s}$$

$$+ C_N h^N \|u\|_{L^2}.$$

Proof In order to obtain Theorem 7.1.5 from Theorem 7.1.1, let $\{\chi_i\}_{i \in \mathbf{i}}$ be a δ-partition for $S_x^* M$ associated to a $(\mathfrak{D}_n, \tau, R(h))$ cover $\{\Lambda_{\rho_i}^\tau(R(h))\}_{i \in \mathbf{i}}$ (see Lemmas 6.2.6 and 6.2.7 for their existence). We need to estimate sums in the right-hand side of

$$\|u\|_{L^\infty(B(x, \frac{1}{16} R(h)))}$$

$$\le \frac{C_n}{\tau^{1/2}} h^{\frac{1-n}{2}} R(h)^{\frac{n-1}{2}} \left(\sum_{i \in \mathcal{I}} \|Op_h(\chi_i)u\|_{L^2} \right.$$

$$\left. + C h^{-1} \|Op_h(\chi_i)(-h^2 \Delta_g - 1)u\|_{L^2} \right)$$

$$+ C C_s h^{-\frac{n}{2}} R(h)^{-1} \|(-h^2 \Delta - 1)u\|_{H_h^s} + C_N h^N \|u\|_{L^2} \qquad (7.14)$$

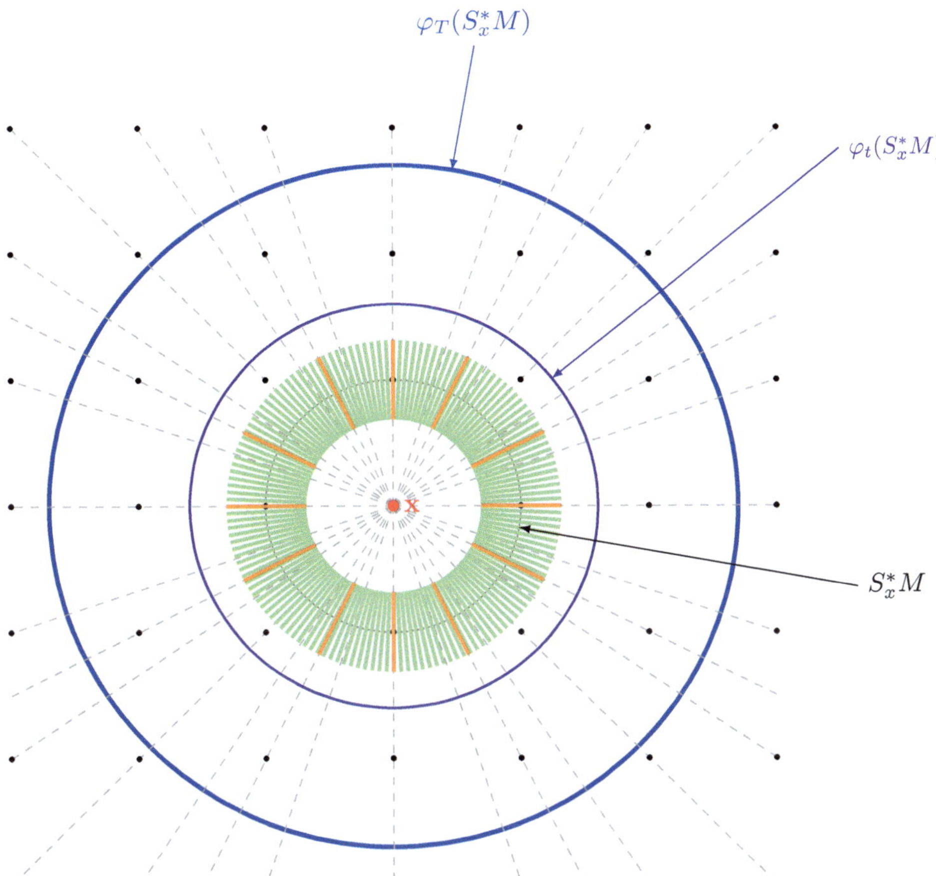

Fig. 7.1 In the figure we illustrate how to cover S_x^*M with "good" tubes (green) and "bad" tubes (orange) for a point x on the square flat torus. The grid represents the integer lattice on the universal cover of the torus. In the figure, there is *only one index* i.e. $\ell = 1$, and we chose $t_\ell = t = 1.6$, $T_\ell = T = 2.7, \tau = 0.2$, and $R = 0.01$. In the figure, the length of the green/orange tubes is $2(\tau + R)$. Note that some of the green tubes *are not* $[3\tau, T]$ non-self looping but are $[t, T]$ non-self looping e.g. the tube at angle $\pi/4$. In practice, to obtain quantitative gains, one needs to work with $T \to \infty$. The figure is drawn for *one* relatively small T because choosing a larger T makes the figure illegible. A tube is "bad" if the geodesic generated by it returns to x in time between t and T. Note, in addition, that t_ℓ must be positive since our tubes have finite, positive width in the flow direction. Also, a set *may* be $[t_0, T]$ self-looping, but not $[\tilde{t}_0, T]$ self-looping for some $\tilde{t}_0 > t_0$ e.g. a neighborhood, $U \setminus V \subset T^*M$, where U is a neighborhood around an unstable hyperbolic closed geodesic in phase space and V is a slightly smaller neighborhood

We first estimate the second term in the sum. Observe that

$$\sum_{i\in\mathcal{I}} \|Op_h(\chi_i)(-h^2\Delta_g - 1)u\|_{L^2} \leq |\mathcal{I}|^{1/2}\Big(\sum_{i\in\mathcal{I}} \|Op_h(\chi_i)(-h^2\Delta_g - 1)u\|_{L^2}^2\Big)^{1/2}$$
$$\leq C_n\mathfrak{D}_n^{1/2}R(h)^{\frac{1-n}{2}}\Big(\sum_{i\in\mathcal{I}} \|Op_h(\chi_i)(-h^2\Delta_g - 1)u\|_{L^2}^2\Big)^{1/2}.$$

Next,

$$\sum_{i\in\mathcal{I}} \|Op_h(\chi_i)(-h^2\Delta_g - 1)u\|_{L^2}^2$$
$$= \Big\langle \sum_{i\in\mathcal{I}} Op_h(\chi_i)^*Op_h(\chi_i)(-h^2\Delta_g - 1)u,\, (-h^2\Delta_g - 1)u\Big\rangle_{L^2}$$

Now, since χ_i is subordinate to a $(\mathfrak{D}_n, \tau, R(h))$ good cover of S_x^*M,

$$\sum_{j\in\mathcal{J}} Op_h(\chi_i)^*Op_h(\chi_i)(-h^2\Delta_g - 1)u$$
$$= Op_h(e)^*Op_h(e)(-h^2\Delta_g - 1)u + O(h^\infty)_{\Psi-\infty})u,$$

for some $e \in S_\delta$ with $\sigma(e) \leq \mathfrak{D}_n^{\frac{1}{2}}$. In particular,

$$\sum_{i\in\mathcal{I}} \|Op_h(\chi_j)(-h^2\Delta_g - 1)u\|_{L^2}^2 \leq \|Op_h(e)(-h^2\Delta - 1)u\|_{L^2}^2 + C_N h^N\|u\|_{L^2}^2,$$

and hence

$$\sum_{i\in\mathcal{I}} \|Op_h(\chi_i)(-h^2\Delta_g - 1)u\|_{L^2} \leq C_n\mathfrak{D}_n R(h)^{\frac{1-n}{2}}\|(-h^2\Delta - 1)u\|_{L^2} + C_N h^N\|u\|_{L^2}.$$
$$(7.15)$$

To estimate the first term in (7.14), we first let $\{\psi_j\}_{j\in\mathcal{J}}$ be a δ-partition of S_x^*M associated to $\{\Lambda_{\rho_j}^\tau(R(h))\}_{j\in\mathcal{J}}$. Then arguing as in (7.8) and (7.9), we obtain

$$\sum_{i\in\mathcal{I}} \|Op_h(\chi_i)u\|_{L^2} \leq C_n\sum_{j\in\mathcal{J}} \|Op_h(\psi_j)u\|_{L^2} + C_N h^N\|u\|_{L^2}. \qquad (7.16)$$

Next, we write

$$\sum_{j\in\mathcal{J}} \|Op_h(\psi_j)u\|_{L^2} = \sum_{j\in\mathcal{B}} \|Op_h(\psi_j)u\|_{L^2} + \sum_{\ell\in\mathcal{L}}\sum_{j\in\mathcal{G}_\ell} \|Op_h(\psi_j)u\|_{L^2}. \qquad (7.17)$$

Observe that

$$R(h)^{\frac{n-1}{2}} \sum_{j \in \mathcal{G}_\ell} \|Op_h(\psi_j)u\|_{L^2} \le R(h)^{\frac{n-1}{2}} |\mathcal{G}_\ell|^{\frac{1}{2}} \Big(\Big\langle \sum_{j \in \mathcal{G}_\ell} Op_h(\psi_j)^* Op_h(\psi_j)u, u \Big\rangle_{L^2} \Big)^{\frac{1}{2}}.$$

(7.18)

Let $a_\ell \in S_\delta^{\mathrm{comp}}$ such that $Op_h(a_\ell) := \sum_{j \in \mathcal{G}_\ell} Op_h(\psi_j)^* Op_h(\psi_j) + O(h^\infty)_{\Psi-\infty}$. Then $\sup |a_\ell| \le (1 + Ch^{1-2\delta})$ and we may apply Lemma 7.1.4 to obtain

$$|\langle Op_h(a_\ell)u, u \rangle| \le \Big(\frac{t_\ell}{T_\ell} + O(h^{1-2\gamma}) \Big) \|u\|_{L^2}^2 + \frac{1}{h^2} t_\ell T_\ell \|(-h^2 \Delta_g - 1)u\|_{L^2}^2.$$

Combining this with (7.14), (7.15), (7.16), (7.17), and (7.18) completes the proof. $\qquad\square$

7.2 Applications to Averages

This section contains applications of the geodesic beam methods to estimates of averages of quasimodes for the Laplacian over submanifolds $H \subset M$. Since $\{x\} \subset M$ is, in particular, a submanifold of codimension n, these results generalize those from Sect. 7.1 to the case of smaller codimension. In fact, the proofs of these theorems (which can be found in [CG21]) are very similar to those of the corresponding theorems in Sect. 7.1. Because of this similarity, we omit the details here, instead indicating the changes which need to be made. The results in this section will later be applied to obtain improvements on estimates for the remainder in Weyl Laws (see Sect. 7.2.1).

The analog of Theorem 7.1.1 is given by:

Theorem 7.2.1 *Let $H \subset M$ be an embedded submanifold of codimension k. There exist $\tau_0 = \tau_0(M, g) > 0$, $R_0 = R_0(M, g) > 0$, $C_n > 0$ depending only on n, so that the following holds.*

Let $0 < \tau \le \tau_0$, $0 \le \delta < \frac{1}{2}$, and $R(h) := \min(h^\delta, R_0)$. Let $\{\chi_j\}_{j \in J}$ be a δ-partition for SN^*H associated to a $(\tau, R(h))$-cover $\{\Lambda_{\rho_j}^\tau(R(h))\}_{j \in \mathcal{J}}$. Let $N > 0$ and $s > \frac{k}{2} - 2$, and $e \in S_\delta$ with

$$\mathrm{supp}(1 - e) \cap \{\rho \in T^*M \,:\, d(\rho, S^*M) > cR(h)\} = \emptyset.$$

Then, there are $h_0 = h_0(M, g, \{\chi_j\}, \delta) > 0$, $C_s > 0$ and $C_N > 0$ with the property that for any $0 < h < h_0$ and $u \in \mathcal{D}'(M)$

$$\Big| \int_H u \, d\sigma_H \Big|$$

$$\le C_n \tau^{-\frac{1}{2}} h^{\frac{1-k}{2}} R(h)^{\frac{n-1}{2}} \sum_{j \in \mathcal{J}} \Big(\|Op_h(\chi_j)u\|_{L^2} + h^{-1} \|Op_h(\chi_j)(-h^2 \Delta_g - 1)u\|_{L^2} \Big)$$

$$+ C_s h^{-1} h^{-\frac{k-1}{2}+1} R(h)^{-1} \|Op_h(e)(-h^2 \Delta_g - 1)u\|_{H_h^s(M)} + C_N h^N \|u\|_{L^2}.$$

(7.19)

Moreover, the constants h_0 and C_N are uniform for χ_j in bounded subsets of S_δ.

Proof Here we present a sketch of changes from the proof of Theorem 7.2.1.

Step 1 (Only conormal directions contribute). We first show that if $a \in S_\delta^{\mathrm{comp}}(T^*M)$ such that

$$d(\mathrm{supp}\, a, N^*H) > c_1 h^\delta, \tag{7.20}$$

then for any $u \in L^2(M)$,

$$\int_H Op_h(a)u\, d\sigma_H = O(h^\infty)\|u\|_{H_h^{-N}}.$$

This follows from the facts that for $b \in S_\delta^{\mathrm{comp}}(T^*H)$ with

$$d(\mathrm{supp}\, b, \{(x,0) \in T^*H\}) > c_2 h^\delta, \tag{7.21}$$

we have for all $v \in L^2(H)$,

$$\int_H Op_h(b)u\, d\sigma_H = O(h^\infty)\|u\|_{H_h^{-N}},$$

and that for a satisfying (7.20), there is $c_2 > 0$ and b satisfying (7.21) such that

$$\|Op_h(b)\gamma_H Op_h(a)\|_{L^2(M)\to H_h^N(H)} \leq C_N h^N, \tag{7.22}$$

where $\gamma_H : C^\infty(M) \to C^\infty(H)$ denotes the restriction map. These two facts follow from elementary integration by parts arguments (see [CG21, Proof of Proposition 3.5]).

Step 2 (Estimates away from S^*M). As in the proof of Theorem 7.1.1, for $a \in S_\delta(T^*M)$ with

$$d(S^*M, \mathrm{supp}\, a) \geq cR(h),$$

we have the estimate

$$\|Op_h(a)u\|_{L^2(H)} \leq CR(h)^{-1}h^{-\frac{k}{2}}\|Op_h(e)(-h^2\Delta_g - 1)u\|_{H_h^s(M)}. \tag{7.23}$$

Step 3 (Geodesic beam estimates for conormal tubes). Next, we use Lemma 6.2.7 to create a δ-good partition of unity $\{\chi_j\}_{j\in\mathcal{J}}$ near SN^*H associated to a $(\mathfrak{D}_N, \tau, r)$ good cover of SN^*H and observe that by (7.22) and (7.23) it remains to estimate

$$\left| \int_H \sum_{j\in\mathcal{J}} Op_h(\chi_j)u\, d\sigma_H \right|.$$

We then apply Lemma 6.1.3, which gives

$$\|\gamma_H Op_h(\chi_j)u\|_{L^1(H)}$$
$$\leq \frac{C_{n,k}}{\tau^{\frac{1}{2}}} h^{\frac{1-k}{2}} R(h)^{\frac{n-1}{2}} \left(\|Op_h(\chi_j)u\|_{L^2} + h^{-1}\|Op_h(\chi_j)(-h^2\Delta_g - 1)u\|_{L^2} + C_N h^N \|u\|_{L^2} \right).$$

$$\square$$

The analog of Theorem 7.1.5 is given below. Its proof is identical to that of Theorem 7.1.5, one simply uses Theorem 7.2.1 in place of Theorem 7.1.1.

Theorem 7.2.2 *Let $H \subset M$ be an embedded submanifold of codimension k, $0 < \delta < \frac{1}{2}$, $s > \frac{k}{2} - 2$, and $N > 0$. There exist positive constants $h_0 = h_0(M, g, \delta)$, $\tau_0 = \tau_0(M, g)$, $R_0 = R_0(M, g)$, and C_n depending only on n, so that for all $0 < \tau \le \tau_0$ and $0 < h < h_0$ the following holds.*

Let $R(h) := \min(h^\delta, R_0)$, and $\{\Lambda^\tau_{\rho_j}(R(h))\}_{j=1}^{N_h}$ be a $(\tau, R(h))$-cover for SN^*H. Let $0 \le \gamma < \frac{1}{2} - \delta$ and suppose there exists a partition of $\{1, \dots, N_h\}$ into $\mathcal{B}$ and $\{\mathcal{G}_\ell\}_{\ell \in \mathcal{L}}$ such that for every $\ell \in \mathcal{L}$ there exist $T_\ell = T_\ell(h) > 0$ and $t_\ell = t_\ell(h) > 0$ with $t_\ell(h) \le T_\ell(h) \le \gamma T_e(h)$ such that

$$\bigcup_{j \in \mathcal{G}_\ell} \Lambda^\tau_{\rho_j}(R(h)) \quad is \quad [t_\ell, T_\ell] \ non\text{-}self \ looping.$$

Then, for all $N > 0$ there exists $C_N = C_N(M, g, N, \tau, \delta) > 0$ so that for $u \in \mathcal{D}'(M)$

$$\left| \int_H u \, d\sigma_H \right| \le C_{n,k} \tau^{-\frac{1}{2}} h^{\frac{1-k}{2}} R(h)^{\frac{n-1}{2}} \left(|\mathcal{B}|^{\frac{1}{2}} \|u\|_{L^2} + \sum_{\ell \in \mathcal{L}} \frac{|\mathcal{G}_\ell|^{\frac{1}{2}} t_\ell^{\frac{1}{2}}}{T_\ell^{\frac{1}{2}}} \|u\|_{L^2} \right.$$

$$\left. + (|\mathcal{G}_\ell| t_\ell T_\ell)^{\frac{1}{2}} h^{-1} \|(-h^2 \Delta_g - 1)u\|_{L^2} \right)$$

$$+ C_{n,k} \tau^{-\frac{1}{2}} h^{-\frac{1-k}{2}} \|(-h^2 \Delta_g - 1)u\|_{L^2}$$

$$+ C_s h^{-\frac{k}{2}} R(h)^{-1} \|(-h^2 \Delta_g - 1)u\|_{H^s_h} + C_N h^N \|u\|_{L^2}.$$

7.2.1   Remarks on the Analog of Theorem 7.2.2 for More General Operators

Theorems 7.2.1 and 7.2.2 hold for more general operators than $(-h^2 \Delta_g - 1)$. Indeed, if one replaces $(-h^2 \Delta_g - 1)$ by any self-adjoint pseudodifferential operator $\mathbf{P}$ satisfying $|\sigma(\mathbf{P})| \ge \frac{1}{C} |\xi|^2 - C$ such that $d\pi_M H_{\sigma(\mathbf{P})}(\rho)$ is transverse to H for all $\rho \in \{\sigma(P) = 0\} \cap N^*H$, and replaces the tubes $\Lambda_{\rho_j}^\tau(R(h))$ by tubes generated by the Hamiltonian flow for $\sigma(\mathbf{P})$ covering $\{\sigma(\mathbf{P}) = 0\} \cap N^*H$. Then the analog of the estimate in Theorem 7.2.1 holds:

$$\left| \int_H u \, d\sigma_H \right| \le C_n \tau^{-\frac{1}{2}} h^{\frac{1-k}{2}} R(h)^{\frac{n-1}{2}} \sum_{j \in \mathcal{J}} \left(\|Op_h(\chi_j)u\|_{L^2} + h^{-1} \|Op_h(\chi_j)\mathbf{P}u\|_{L^2} \right)$$

$$+ C_s h^{-1} h^{-\frac{k-1}{2}+1} R(h)^{-1} \|Op_h(e)\mathbf{P}u\|_{H^s_h M)} + C_N h^N \|u\|_{L^2}, \qquad (7.24)$$

where $d(\mathrm{supp}\, e, \{\sigma(\mathbf{P}) = 0\}) \ge c R(h)$.

The analog of Theorem 7.2.2 also holds

$$\left| \int_H u\, d\sigma_H \right| \le C_{n,k}\tau^{-\frac{1}{2}} h^{\frac{1-k}{2}} R(h)^{\frac{n-1}{2}} \left(|\mathcal{B}|^{\frac{1}{2}} \|u\|_{L^2} + \sum_{\ell \in \mathcal{L}} \frac{|\mathcal{G}_\ell|^{\frac{1}{2}} t_\ell^{\frac{1}{2}}}{T_\ell^{\frac{1}{2}}} \|u\|_{L^2} \right.$$

$$\left. + (|\mathcal{G}_\ell| t_\ell T_\ell)^{\frac{1}{2}} h^{-1} \|Pu\|_{L^2} \right)$$

$$+ C_{n,k}\tau^{-\frac{1}{2}} h^{-\frac{1-k}{2}} \|Pu\|_{L^2} + C_s h^{-\frac{k}{2}} R(h)^{-1} \|Pu\|_{H_h^s} + C_N h^N \|u\|_{L^2}. \qquad (7.25)$$

For the Weyl Law result, presented in Sect. 2.2.3, we will want to apply these estimates
with M replaced by $M \times M$, $k = n$, the hypersurface $H = \mathbf{\Delta}$ given by the diagonal

$$\mathbf{\Delta} := \{(x, x) \in M \times M : x \in M\},$$

and the function,

$$u(x, y) = \mathbb{1}_{[1-t,1]}(-h^2 \Delta_g)(x, y), \qquad x, y \in M.$$

The operator here will be $\mathbf{P} = (-h^2 \Delta_g - 1) \otimes Id$ so that for $|t| \le \frac{1}{2}$,

$$\|Pu\|_{H_h^s} \le C_s |t| \left(\#\{\lambda_j \in \mathrm{Spec}(-h^2 \Delta_g) : \lambda_j \in [1-t, 1]\} \right)^{\frac{1}{2}}.$$

Note, however, that

$$|\sigma(\mathbf{P})(x, \xi, y, \eta)| \le C(|\xi|^2 + 1),$$

and hence does not satisfy the required ellipticity property:

$$|\sigma(\mathbf{P})(x, \xi, y, \eta)| \ge \tfrac{1}{C} |(\xi, \eta)|^2 - C.$$

To get around this, let $\psi \in S^0(T^*M \times M)$ with $\psi \equiv 1$ on

$$\{(x, \xi, y, \eta) \in T^*M \times T^*M : \tfrac{1}{2}|\xi|_g < |\eta|_g < 2|\xi|_g \text{ or } |\xi|_g \le 3\}$$

and

$$\mathrm{supp}\,\psi \subset \{(x, \xi, y, \eta) \in T^*M \times T^*M : \tfrac{1}{3}|\xi|_g < |\eta|_g < 3|\xi|_g \text{ or } |\xi|_g \le 4\}.$$

We then set

$$\tilde{\mathbf{P}} := \mathbf{P} + Op_h\left((|\xi|_g^2 + |\eta|_g^2)(1 - \psi) \right),$$

and apply the estimate with $\tilde{\mathbf{P}}$. We then observe that with u the kernel of $\mathbb{1}_{[1-t,1]}(-h^2 \Delta_g)$,
we have $\mathrm{WF}_h(u) \subset \{\psi \equiv 1\}$, and hence

$$\|\tilde{\mathbf{P}}u\|_{H_h^s} = \|Pu\|_{H_h^s} + C_N h^N \|u\|_{H_h^{-N}}.$$

Thus, we obtain by (7.24) that

$$\left| \int_\Delta u \, d\sigma_\Delta \right| \leq C_n \tau^{-\frac{1}{2}} h^{\frac{1-n}{2}} R(h)^{\frac{n-1}{2}} \sum_{j \in \mathcal{J}} \left(\|Op_h(\chi_j)u\|_{L^2} + h^{-1} \|Op_h(\chi_j)\mathbf{P}u\|_{L^2} \right)$$

$$+ C_s h^{-1} h^{-\frac{n-1}{2}+1} R(h)^{-1} \|Op_h(e)\mathbf{P}u\|_{H_h^s(M)} + C_N h^N \|u\|_{L^2},$$

$$(7.26)$$

where $d(\operatorname{supp} e, S^*M \times T^*M) \geq cR(h)$ and, by (7.25), that

$$\left| \int_\Delta u \, d\sigma_\Delta \right| \leq C_n \tau^{-\frac{1}{2}} h^{\frac{1-n}{2}} R(h)^{\frac{n-1}{2}} \left(|\mathcal{B}|^{\frac{1}{2}} \|u\|_{L^2} + \sum_{\ell \in \mathcal{L}} \frac{|\mathcal{G}_\ell|^{\frac{1}{2}} t_\ell^{\frac{1}{2}}}{T_\ell^{\frac{1}{2}}} \|u\|_{L^2} \right.$$

$$\left. + (|\mathcal{G}_\ell| t_\ell T_\ell)^{\frac{1}{2}} h^{-1} \|\mathbf{P}u\|_{L^2} \right)$$

$$+ C_n \tau^{-\frac{1}{2}} h^{-\frac{1-n}{2}} \|\mathbf{P}u\|_{L^2} + C_s h^{-\frac{n}{2}} R(h)^{-1} \|\mathbf{P}u\|_{H_h^s} + C_N h^N \|u\|_{L^2}. \quad (7.27)$$

7.3 Applications to L^p Norms

Theorem 7.3.1 below controls $\|u\|_{L^p(U)}$ using an assumption on the maximal volume of long geodesics joining any two given points in U. For our proof, it is necessary to control the number points in U where the L^∞ norm of u can be large. This is a very delicate and technical part of the argument, as the points in question may be approaching one another at rates $\sim h^\delta$ as $h \to 0^+$, with $0 < \delta < \frac{1}{2}$.

Theorem 7.3.1 *There exists $\tau_M > 0$ such that for all $p > p_c$ and $\varepsilon_0 > 0$ the following holds. Let $U \subset M$, $0 < \delta_1 < \delta_2 < \frac{1}{2}$ and let $h^{\delta_2} \leq R(h) \leq h^{\delta_1}$ for all $h > 0$. Let $1 \leq T(h) \leq (1 - 2\delta_2)T_e(h)$ and let $t_0 > 0$ be h-independent. Let $\{T_j\}_{j \in \mathcal{J}}$ be a $(\tau, R(h))$ cover for S^*M for some $0 < \tau < \tau_M$.*

Suppose that for any pair of points $x_1, x_2 \in U$, the tubes over x_1 can be partitioned into a disjoint union $\mathcal{J}_{x_1} = \mathcal{B}_{x_1,x_2} \sqcup \mathcal{G}_{x_1,x_2}$ where

$$\bigcup_{j \in \mathcal{G}_{x_1,x_2}} \varphi_t(T_j) \cap S^*_{B(x_2,R(h))}M = \emptyset, \qquad t \in [t_0, T(h)].$$

Then, there are $h_0 > 0$ and $C > 0$ so that for all $u \in \mathcal{D}'(M)$, and $0 < h < h_0$,

$$\|u\|_{L^p(U)} \leq C h^{-\delta(p)} \left(\frac{\sqrt{t_0}}{\sqrt{T(h)}} + \left[\sup_{x_1,x_2 \in U} |\mathcal{B}_{x_1,x_2}| R(h)^{n-1} \right]^{\frac{1}{6+\varepsilon_0}(1-\frac{p_c}{p})} \right)$$

$$\times \left(\|u\|_{L^2} + \frac{T(h)}{h} \|(-h^2 \Delta_g - 1)u\|_{H_h^{\frac{n-3}{2}-\frac{n}{p}}} \right). \quad (7.28)$$

In order to interpret (7.28), note that we think of the tubes $\mathcal{G}_{x_1,x_2}$ and $\mathcal{B}_{x_1,x_2}$ as respectively good (or non- looping) and bad (or looping) tubes. Then, observe that

$$|\mathcal{B}_{x_1,x_2}|R(h)^{n-1} \sim \mathrm{vol}\left(\bigcup_{j\in\mathcal{B}_{x_1,x_2}} \mathcal{T}_j \cap S^*_{x_1}M \right),$$

and $\bigcup_{j\in\mathcal{B}_{x_1,x_2}} \mathcal{T}_j$ is the set of points over x_1 which may loop through x_2 in time $T(h)$. Therefore, if the volume of points in $S^*_{x_1}M$ looping through x_2 is bounded by $T(h)^{-(3+\varepsilon_0)(1-\frac{p_c}{p})^{-1}}$, (7.28) provides $T(h)^{-\frac{1}{2}}$ improvements over the standard L^p bounds. We expect these non-looping type assumptions to be valid on generic manifolds.

7.3.1 Discussion of the Proof of Theorem 7.3.1

Our method for proving Theorem 7.3.1 differs from the standard approaches for treating L^p norms in two major ways. It hinges on adapting the geodesic beam techniques constructed in Chap. 6, and on the development of a new second-microlocal calculus. Because the full proof is technical, we do not reproduce it here. We instead make a detailed outline of the ideas, always assuming that u is a genuine eigenfunction of the Laplacian and hence that we need only concern ourselves with the parts of u microlocalized to an $R(h)$ neighborhood of S^*M. When u is only a quasimode, estimating the parts microlocally away from S^*M involves only a relatively simple elliptic parametrix construction.

7.3.1.1 Step 1 (Setting up the Cover by Tubes)

In order to estimate the full L^p norm of u, it is enough to understand the L^p norm of u coming from a single open, h-independent subset of S^*M. Therefore, we start by creating an appropriate microlocal partition of unity over such a set. We first fix a smooth hypersurface $H \subset M$, and choose Fermi normal coordinates $x = (x_1, x')$ in a neighborhood of $H = \{x_1 = 0\}$. We write $(\xi_1, \xi') \in T^*_x M$ for the dual coordinates. Let

$$\Sigma_H := \left\{ (x,\xi) \in S^*_H M : \; |\xi_1| \geq \tfrac{1}{2} \right\} \tag{7.29}$$

Since, $\Lambda^\tau_{\Sigma_H}(0)$ contains an open subset of S^*M for any $\tau > 0$, it is enough to estimate the L^p norm of Xu where X is microlocally the identity on $\Lambda^\tau_{\Sigma_H}(\tfrac{1}{2}R(h))$.

Our first goal is to construct such a localizer X made of geodesic beam cutoffs. To do this, we construct a $(\mathfrak{D}_n, \tau, R(h))$ good cover of Σ_H, $\{\mathcal{T}_j\}_{j\in\mathcal{J}}$, i.e. a cover by tubes of length τ and radius $R(h)$. Then, with $\{\chi_{\mathcal{T}_j}\}_{j\in\mathcal{J}}$ a good partition associated to $\{\mathcal{T}_j\}_{j\in\mathcal{J}}$,

$$X := \sum_{j\in\mathcal{J}} Op_h(\chi_{\mathcal{T}_j})$$

is microlocally the identity on $\Lambda^\tau_{\Sigma_H}(\tfrac{1}{2}R(h))$.

We have now reduced estimating the L^p norm of u to that estimating that of $v := Xu$.

7.3.1.2 Step 2 (Filtering Tubes by L^∞-Weight on Shrinking Balls)

For each $k \geq -1$, we consider separately the tubes $\{T_j\}_{j \in A_k}$ which satisfy

$$A_k = \left\{ j \in \mathcal{J} : \frac{1}{2^{k+1}} \|u\|_{L^2} \leq \|Op_h(\chi_{T_j})u\|_{L^2} \leq \frac{1}{2^k} \|u\|_{L^2} \right\}. \tag{7.30}$$

We then let

$$w_k := \sum_{j \in A_k} Op_h(\chi_{T_j})u$$

be the contribution to v from these tubes. One can check by almost orthogonality of $Op_h(\chi_{T_j})$ that

$$|A_k| \leq C_n 2^{2k}. \tag{7.31}$$

In order to understand the L^p norm of w_k, we next decompose the manifold into a $\mathfrak{D}_n$ good cover $\{B(x_\alpha, R(h))\}_{\alpha \in \mathrm{i}}$ by balls of radius $R(h)$. By doing this, we are able to think of the L^p norm of a function on M as the L^p norm of a function on a disjoint union of balls of radius $R(h)$. To keep track of these tubes, we put for each $k \in \mathbb{Z}_+$ and $\alpha \in \mathrm{i}$

$$A_k(\alpha) := \{j \in A_k : \pi_M(T_j) \cap B(x_\alpha, 2R(h)) \neq \emptyset\},$$

where $\pi_M : T^*M \to M$ is the standard projection. The indices in A_k are those that correspond to tubes with mass comparable to $\frac{1}{2^k}\|u\|_{L^2}$, while indices in $A_k(\alpha)$ correspond to tubes of mass $\frac{1}{2^k}\|u\|_{L^2}$ that run over the ball $B(x_\alpha, 2R(h))$.

Finally, we need to keep track of the L^∞ norms of w_k on the balls in i. To this end, we define for $m \in \mathbb{Z}$

$$\mathcal{I}_{k,m} := \left\{ \alpha \in \mathcal{I} : 2^{m-1} \leq h^{\frac{n-1}{2}} R(h)^{\frac{1-n}{2}} 2^k \frac{\|w_k\|_{L^\infty(B(x_\alpha, R(h)))}}{\|u\|_{L^2}} \leq 2^m \right\}. \tag{7.32}$$

By an application of the geodesic beam estimate on L^∞ norms, Theorem 7.1.1, one has

$$c_M 2^m \leq |A_k(\alpha)| \qquad \alpha \in \mathrm{i}_{k,m}. \tag{7.33}$$

We next write

$$w_{k,m} := \sum_{j \in A_{k,m}} Op_h(\chi_{T_j})u, \qquad A_{k,m} := \bigcup_{\alpha \in \mathcal{I}_{k,m}} A_k(\alpha). \tag{7.34}$$

for the part of w_k which contributes mass to a point in $\mathcal{I}_{k,m}$ i.e. a ball with L^∞ norm $\sim 2^m \|u\|_{L^2} h^{\frac{1-n}{2}} R(h)^{\frac{n-1}{2}} 2^{-k}$.

7.3.1.3 Step 3 (Splitting of w_k into High and Low L^∞ Pieces)

We have decomposed w_k by its L^∞ norm so that we may treat points with high L^∞ norm separately from those with low L^∞ norm. (Here, the meaning of high and low will be explained below.) Those points with low L^∞ will be treated using a relatively simple interpolation argument while, for those m corresponding to a high L^∞ norm we will make a detailed analysis for each m.

Before moving on to this analysis, we observe two facts about $\mathcal{A}_{k,m}$ which come directly from the geodesic beam estimate (7.33), which in turn is an easy corollary of Theorem 7.1.1. First, by (7.33) and (7.31), we have

$$2^m \geq c_n 2^{2k} \quad \Rightarrow \quad \mathcal{A}_{k,m} = \emptyset.$$

Second, since at most $c_0 R(h)^{1-n}$ tubes $\mathcal{T}_j$ pass over any give ball $B(x_\alpha, R(h))$, we have, again using (7.33)

$$2^m \geq c_0 R(h)^{1-n} \quad \Rightarrow \quad \mathcal{A}_{k,m} = \emptyset.$$

We define $m_{2,k}$ by

$$2^{m_{2,k}} := \min(c_n 2^{2k}, c_0 R(h)^{1-d}).$$

We now move to the splitting of w_k into 'high' and 'low' L^∞ pieces. First, observe that for $w_{k,m}$ defined as in (7.34),

$$\|w_k\|_{L^p}^p \leq \mathfrak{D}_n \sum_{m=-\infty}^{m_{2,k}} \|w_k\|_{L^p(U_{k,m})}^p = \mathfrak{D}_n \sum_{m=-\infty}^{m_{2,k}} \|w_{k,m}\|_{L^p(U_{k,m})}^p + O(h^\infty \|u\|_{L^2}),$$

$$(7.35)$$

where

$$U_{k,m} := \bigcup_{\alpha \in i_{k,m}} B(x_\alpha, R(h)). \qquad (7.36)$$

The split between 'high' and 'low' L^∞ norms will be determined for each k by $m > m_{1,k}$ or $m \leq m_{1,k}$, where

$$2^{m_{1,k}} := \frac{2^k R(h)^{\frac{1-n}{2}}}{T(h)^N},$$

for some large N. In fact, it suffices to take $N > \frac{1}{2}(1 - \frac{p_c}{p})^{-1}$.

7.3.1.4 Step 4 (Control of the Low L^∞ Term)

From (7.35), we have for each $k \geq -1$,

$$\|w_k\|_{L^p}^p \leq \mathfrak{D}_n \sum_{m=-\infty}^{\min(m_{1,k}, m_{2,k})} \|w_{k,m}\|_{L^p(U_{k,m})}^p$$

$$+ \mathfrak{D}_n \sum_{m=m_{1,k}+1}^{m_{2,k}} \|w_{k,m}\|_{L^p(U_{k,m})}^p + O(h^\infty \|u\|_{L^2}), \qquad (7.37)$$

Note that the second sum in (7.37) is empty if $m_{1,k} \geq m_{2,k}$.

We now estimate the 'low' L^∞ term using a simple interpolation argument. Indeed, interpolating between L^{p_c} and L^∞, we obtain

$$\sum_{m=-\infty}^{\min(m_{1,k},m_{2,k})} \|w_{k,m}\|^p_{L^p(U_{k,m})} \leq C \sum_{m=-\infty}^{\min(m_{1,k},m_{2,k})} \|w_{k,m}\|^{p-p_c}_{L^\infty(U_{k,m})} \|w_{k,m}\|^{p_c}_{L^{p_c}(U_{k,m})} \tag{7.38}$$

We then bound the right-hand side of (7.38) by the L^2 norm of u, using the definition (7.32) of $\mathcal{I}_{km}$, the fact that $\frac{1-n}{2}(p - p_c)-1 = -p\delta(p)$, and that

$$\|w_{k,m}\|_{L^{p_c}(U_{k,m})} \leq h^{-\frac{1}{p_c}} \|u\|_{L^2}.$$

Summing in k, we obtain

$$\sum_{k\geq -1} \left(\sum_{m=-\infty}^{\min(m_{1,k},m_{2,k})} \|w_{k,m}\|^p_{L^p(U_{k,m})} \right)^{\frac{1}{p}} \leq C h^{-\delta(p)} \frac{N \log T(h)}{T(h)^{N(1-\frac{p_c}{p})}} \|u\|_{L^2}. \tag{7.39}$$

7.3.1.5 Step 5 (Sharing of Mass for the High L^∞ Term)

It thus remains to understand the high L^∞ piece. The first crucial substep will be to estimate *the number of balls B on which the L^∞ norm of w_k can be large* (i.e. close to extremal). In particular, our goal is to describe the proof of the following estimate.

Lemma 7.3.2 *Let $\frac{1}{2}(\delta + 1) < \rho < 1$. There exists $C > 0$ so that for every $k \geq -1$ and $m \in \mathbb{Z}$ the following holds. If*

$$|\mathcal{A}_{k,m}| \leq C\, 2^{2m} R(h)^{n-1} \left(h^{\rho-\frac{1}{2}} R(h)^{-\frac{1}{2}} \right)^{-\frac{2n(n-1)}{3n+1}},$$

then

$$|\mathcal{I}_{k,m}| \leq C |\mathcal{A}_{k,m}| 2^{-2m} R(h)^{1-n}. \tag{7.40}$$

The idea behind the proof of Lemma 7.3.2 is to understand how much L^2 mass can be effectively shared along short geodesics joining two nearby points in such a way as to produce large L^∞ norm at both points. That is, if x_α and x_β are nearby points on M, and if $|u(x_\alpha)|$ and $|u(x_\beta)|$ are near extremal, how much total L^2 mass must the tubes over x_α and x_β carry?

When considering a point $y \in M$ where the value of an eigenfunction is large, it is natural to try to understand the set in phase space which can contribute energy to the point. If one uses *unstructured* localizers, i.e. ones which are not compatible with the equation, it would be natural to localize to $T_y^* M$. In this case, the localizers take the form

$$\chi(h^{-1}d(x, y)),$$

where $\chi \in C_c^\infty(\mathbb{R})$ is one near 0. Clearly,

$$|u(y)| = |\chi(h^{-1}d(x, y))u(y)|,$$

for any function u. However, when considering eigenfunctions, it is more useful to use *structured* localizers, i.e. ones compatible with the equation as described in Chap. 1. In this case, the phase space set which contributes energy to y is $\bigcup_{|t| < \frac{1}{2} \mathrm{inj}(M)} \varphi_t(T_y^* M)$. Since the energy of u is localized near $|\xi|_g = 1$, in fact, we can localize to

$$\Gamma_y := \bigcup_{|t| < \frac{1}{2} \mathrm{inj}(M)} \varphi_t(\Omega_y), \qquad \Omega_y := \{\xi \in T_y^* M : |1 - |\xi|_{g(y)}| < \delta\}.$$

Because Γ_y is co-isotropic, one can, in principle, construct a localizer X_y h-close to Γ_y. Furthermore, in order to preserve the value of u at y, we would also like

$$\chi(h^{-1}d(x, y))Op_h(\chi(\delta^{-1}(|\xi|_g - 1)))X_y$$
$$= \chi(h^{-1}d(x, y))Op_h(\chi(\delta^{-1}(|\xi|_g^2 - 1))) + O(h^\infty)_{\Psi - \infty}.$$

In fact, we construct X_y so that

$$\begin{cases} \|X_y\|_{L^2 \to L^2} \leq 1, \\ \chi_{h,y} X_y = \chi_{h,y} + O(h^\infty)_{\Psi - \infty}, \\ \mathrm{WF}_h([P, X_y]) \cap \{(x, \xi) : x \in B(y, \frac{1}{2}\mathrm{inj}\, M), \ \xi \in \Omega_x\} = \emptyset, \end{cases} \qquad (7.41)$$

where

$$\chi_{h,y} := \chi(h^{-1}d(x, y))Op_h(\chi(\delta^{-1}(|\xi|_g - 1))).$$

To demonstrate how we understand 'sharing' of mass between two points, suppose u is maximal at $\{y_j\}_{j \in \mathcal{J}}$ i.e.

$$|u(y_j)| \geq ah^{\frac{1-n}{2}}\|u\|_{L^2}.$$

Then, since all of the energy seen at y_j is comes from Γ_{y_j}, one can show, as in [CG20a, (3.29)], by using the standard L^∞ estimate (Theorem 5.3.1) that

$$\|X_{y_j}u\|_{L^2} \geq b\|u\|_{L^2}.$$

If there were no 'sharing' between the Γ_{y_j}, we would have $X_{y_j}^* X_{y_k} = 0$ for $y \neq k$, and hence

$$|\mathcal{J}|b^2\|u\|_{L^2}^2 = \sum_{j \in \mathcal{J}} \|X_{y_j}u\|_{L^2}^2 = \|\sum_{j \in \mathcal{J}} X_{y_j}u\|_{L^2}^2 \leq \sup_{j \in \mathcal{J}} \|X_{y_j}\|^2\|u\|_{L^2}^2 \leq \|u\|_{L^2}^2. \quad (7.42)$$

In particular, $|\mathcal{J}| \leq b^{-2}$. In practice, some mass can be shared between the cutoffs X_{y_j}. We quantify this sharing by obtaining a bound in the form

$$\|X^*_{y_j} X_{y_k}\|_{L^2 \to L^2} \leq C h^{\frac{n-1}{2}} d(y_j, y_k)^{\frac{1-n}{2}}. \tag{7.43}$$

and using the Cotlar–Stein Lemma to handle the extra terms which appear in (7.42).

The fact (7.43) holds comes from an uncertainty type principle which is valid since localization to the intersection of Γ_{y_j} and Γ_{y_k} implies localization to two non-symplectically orthogonal directions (see Fig. 7.2).

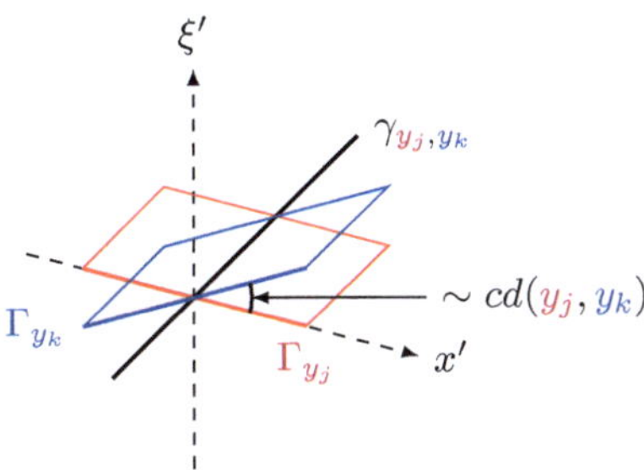

Fig. 7.2 A pictorial representation of the co-isotropics involved in (7.43) where γ_{y_j, y_k} is the geodesic from y_j to y_k. Localization to both Γ_{y_j} and Γ_{y_k} implies localization in the non-symplectically orthogonal directions, x' and ξ'. The uncertainty principle then rules this behavior out

To construct the localizer X_y, one needs a second microlocal calculus associated to a Lagrangian foliation L over a co-isotropic submanifold $\Gamma \subset T^*M$. This calculus allows for simultaneous localization along a leaf of L and along Γ. The calculus, which is developed in [CG20a, Sect. 5], can be seen as an interpolation between those in [DZ16, SZ99]. In particular, for each y we use the calculus associated to the co-isotropic Γ_y with Lagrangian foliation given by

$$L_y := \{L_{t,y}\}_t, \qquad L_{t,y} := \varphi_t(\Omega_y).$$

It is then the incompatibility between the calculi coming from two nearby points which allows us to prove the uncertainty principle type of estimate (7.43).

Remark 7.3.3 In practice, it is simpler to build a second microlocal calculus which allows for h^ρ-localization to Γ and L for some $0 < \rho < 1$, rather than h-localization. This results in a corresponding change in (7.43) but is sufficient to prove Lemma 7.3.2.

7.3.1.6 Step 6 (Control of the High L^∞ Mass Term)

Once the number of balls with high L^∞ norm is understood via Lemma 7.3.2, it remains to employ the non-looping techniques from [CG21] where the L^2 mass on a collection of tubes is estimated using its non-looping time.

To do this, we split

$$\mathcal{A}_{k,m} = \mathcal{G}_{k,m} \sqcup \mathcal{B}_{k,m},$$

where the set of 'good' tubes $\bigcup_{j \in \mathcal{G}_{k,m}} \mathcal{T}_j$ is $[t_0, T]$ non-self looping and the number of 'bad' tubes $|\mathcal{B}_{k,m}|$ is small. To do this, let

$$\mathcal{B}(\alpha, \beta) := \left\{ j \in \bigcup_k \mathcal{A}_k(\alpha) \ : \ \bigcup_{t=t_0}^{T} \varphi_t(\mathcal{T}_j) \cap S_{B(x_\beta, 2R(h))}^* M \neq \emptyset \right\}. \tag{7.44}$$

Then, we define

$$\mathcal{B}_{k,m} := \bigcup_{\alpha, \beta \in \mathcal{I}_{k,m}} \mathcal{B}(\alpha, \beta) \cap \mathcal{A}_k(\alpha).$$

Let $\mathcal{G}_{k,m} := \mathcal{A}_{k,m} \setminus \mathcal{B}_{k,m}$. Then, by construction, $\bigcup_{j \in \mathcal{G}_{k,m}} \mathcal{T}_j$ is $[t_0, T]$ non-self looping and we have

$$|\mathcal{B}_{k,m}| \leq c |\mathcal{I}_{k,m}|^2 |\mathcal{B}| \tag{7.45}$$

for some $c > 0$, where

$$|\mathcal{B}| := \sup\{|\mathcal{B}(\alpha, \beta)| \ : \ \alpha, \beta \in \mathcal{I}\}, \tag{7.46}$$

That is, $|\mathcal{B}|$ is the maximum number of loops of length in $[t_0, T]$ joining any two points in M.

Then, define

$$w_{k,m}^{\mathcal{G}} := \sum_{j \in \mathcal{G}_{k,m}} Op_h(\tilde{\chi}_{\mathcal{T}_j}) Op_h(\psi) u, \qquad w_{k,m}^{\mathcal{B}} := \sum_{j \in \mathcal{B}_{k,m}} Op_h(\tilde{\chi}_{\mathcal{T}_j}) Op_h(\psi) u, \tag{7.47}$$

and we estimate

$$\left(\sum_{m=m_{1,k}+1}^{m_{2,k}} \|w_{k,m}\|_{L^p(U_{k,m})}^p \right)^{\frac{1}{p}} \leq \left(\sum_{m=m_{1,k}+1}^{m_{2,k}} \|w_{k,m}^{\mathcal{G}}\|_{L^p(U_{k,m})}^p \right)^{\frac{1}{p}}$$
$$+ \left(\sum_{m=m_{1,k}+1}^{m_{2,k}} \|w_{k,m}^{\mathcal{B}}\|_{L^p(U_{k,m})}^p \right)^{\frac{1}{p}}. \tag{7.48}$$

To complete the proof, we estimate the first and second terms on the right-hand side of (7.48) separately. We start by estimating the 'bad' term.

To estimate the 'bad' term, we use estimates on three quantities: (1) The number of bad tubes which pass over a point in $i_{k,m}$ is estimated using a dynamical hypothesis. This is then combined together geodesic beam estimate (Theorem 7.1.1)). The total number of bad tubes (7.45) is estimated using a dynamical hypothesis and the estimate (3) on the number of points in $i_{k,m}$, combing from the previous step (Lemma 7.3.2).

First, we note that for each point in $i_{k,m}$ there are at most $c|i_{k,m}||\mathcal{B}|$ tubes in $\mathcal{B}_{k,m}$ touching it. Therefore, we may apply the geodesic beam estimate on L^∞ norms (Theorem 7.1.1) to obtain $C > 0$ such that

$$\|w_{k,m}^{\mathcal{B}}\|_{L^\infty(U_{k,m})} \leq C h^{\frac{1-n}{2}} R(h)^{\frac{n-1}{2}} |\mathcal{I}_{k,m}| |\mathcal{B}| 2^{-k} \|u\|_{L^2}. \tag{7.49}$$

Using (7.49) and interpolating between L^∞ and L^{p_c} we obtain

$$\|w_{k,m}^{\mathcal{B}}\|_{L^p(U_{k,m})}^p \le Ch^{-p\delta(p)}\left(R(h)^{\frac{n-1}{2}}|\mathcal{I}_{k,m}||\mathcal{B}|2^{-k}\|u\|_{L^2}\right)^{p-p_c}\|w_{k,m}^{\mathcal{B}}\|_{L^2(U_{k,m})}^{p_c}. \quad (7.50)$$

In addition, combining (7.30) with (7.45) yields

$$\|w_{k,m}^{\mathcal{B}}\|_{L^2(U_{k,m})} \le C|\mathcal{B}_{k,m}|^{\frac{1}{2}}2^{-k}\|u\|_{L^2} \le C2^{-k}|\mathcal{I}_{k,m}||\mathcal{B}|^{\frac{1}{2}}\|u\|_{L^2}, \quad (7.51)$$

Finally, by (7.40),

$$|i_{k,m}| \le CR(h)^{1-n}2^{-2m}|\mathcal{A}_{k,m}| \le CR(h)^{1-n}2^{2k-2m}, \quad (7.52)$$

and hence, combining (7.50), (7.51), (7.52) and the definition of $m_{j,k}$, $j = 1, 2$, we obtain

$$\sum_{k\ge-1}\left(\sum_{m=m_{1,k}+1}^{m_{2,k}}\|w_{k,m}^{\mathcal{B}}\|_{L^p(U_{k,m})}^p\right)^{\frac{1}{p}} \le Ch^{-\delta(p)}(R(h)^{n-1}|\mathcal{B}|)^{1-\frac{p_c}{2p}}\mathbf{T}(h)^{3N}\|u\|_{L^2}. \quad (7.53)$$

To estimate the 'good' term, we use the estimates on the number of bad tubes, coming from the dynamical hypothesis to show that the L^∞ norm of $w_{k,m}^{\mathcal{G}}$ on any ball $B(x_\alpha, R(h))$ with $x_\alpha \in \mathcal{I}_{k,m}$ is similar to that of $w_{k,m}$. Then, we show that the number of tubes in $\mathcal{A}_{k,m}$ is smaller than the standard estimate (see (7.31)) by a factor of $1/\mathbf{T}(h)$. Together with interpolation and the L^{p_c} estimate, this will complete the proof.

Using (7.49), Lemma 7.3.2, the dynamical hypothesis on $|\mathcal{B}|$, and the definition of $\mathcal{I}_{k,m}$ one obtains

$$\|w_{k,m}^{\mathcal{G}}\|_{L^\infty(B(x_\alpha, R(h)))} \le C2^{m-k}h^{\frac{1-n}{2}}R(h)^{\frac{n-1}{2}}\|u\|_{L^2}; \quad (7.54)$$

i.e. that the L^∞ norm of the 'good' part of $w_{k,m}$ is essentially unchanged from that of $w_{k,m}$ itself.

Next, we need an estimate on the number of tubes in $\mathcal{A}_{k,m}$. Here, we use the non-looping property of the good tubes. We prove in [CG20a, Lemma 3.4] that given $k \in \mathbb{Z}$, $k \ge -1$, and $t_0 > 1$, if $\mathcal{G} \subset \mathcal{A}_k$ is such that

$$\bigcup_{j\in\mathcal{G}} \mathcal{T}_j \quad \text{is } [t_0, \mathbf{T}] \text{ non-self looping,}$$

then there exists a constant $C_n > 0$, depending only on n, such that

$$|\mathcal{G}| \le \frac{C_n t_0}{\mathbf{T}}2^{2k}.$$

In particular, since $\mathcal{A}_{k,m}\setminus\mathcal{B}_{k,m}$ is $[t_0, T]$ non-self looping,

$$|\mathcal{A}_{k,m}| - C|i_{k,m}|^2|\mathcal{B}| \le |\mathcal{A}_{k,m}| - |\mathcal{B}_{k,m}| \le \frac{C_n t_0}{\mathbf{T}}2^{2k}.$$

Using once again, the estimate from Lemma 7.3.2 on $|i_{k,m}|$ and the control on $|\mathcal{B}|$ from the dynamical hypothesis, together with (7.33) we obtain for any m where $|i_{k,m}| \neq 0$,

$$c_M 2^m \leq |\mathcal{A}_{k,m}| \leq \frac{C_n t_0}{\mathbf{T}} 2^{2k}. \tag{7.55}$$

The estimate (7.55) has two important consequences. First, defining

$$2^{m_{3,k}} := \frac{C_n t_0}{\mathbf{T}} 2^{2k},$$

we have $|i_{k,m}| = \emptyset$ for $m > m_{3,k}$. Second, we have

$$\|w_{k,m}^{\mathcal{G}}\|_{L^2}^2 \leq \frac{C_n t_0}{\mathbf{T}} \|u\|_{L^2}.$$

Using these estimates, interpolating between L^∞ and L^{p_c} as before, and summing in k, we obtain

$$\sum_{k=-1}^{\infty} \left(\sum_{m=m_{1,k}+1}^{m_{2,k}} \|w_{k,m}^{\mathcal{G}}\|_{L^p(U_m)}^p \right)^{\frac{1}{p}} = \sum_{k=-1}^{\infty} \left(\sum_{m=m_{1,k}+1}^{\min(m_{2,k},m_{3,k})} \|w_{k,m}^{\mathcal{G}}\|_{L^p(U_m)}^p \right)^{\frac{1}{p}} \tag{7.56}$$

$$\leq C h^{-\delta(p)} \frac{t_0^{\frac{1}{2}}}{\mathbf{T}^{\frac{1}{2}}} \|u\|_{L^2}.$$

These completes the proof since we have now estimated the L^p norm of all pieces as claimed. In particular, the estimates (7.48), (7.39), (7.53), and (7.56) imply the theorem.

7.4   Applications to Weyl Asymptotics

In this section, we discuss the application of geodesic beams to asymptotics for the spectral projector for the Laplacian. We will give a detailed proof of Theorem 2.2.5. The proof of Theorem 2.2.7 follows from a similar proof for which we refer the reader to [CG20b].

7.4.1   Outline of the Proof of Theorem 2.2.5

In order to prove Theorem 2.2.5, there are two important estimates. The first compares the counting function to a version smoothed at small scales, and the second compares the counting function smoothed at small scales to one smoothed at larger scales.

Let $\hat{\rho} \in C_c^\infty(\mathbb{R})$ with $\hat{\rho} \equiv 1$ on $[-1, 1]$, and put

$$\rho_{h,T}(s) := \frac{T}{h} \rho\left(\frac{T}{h} s\right).$$

Then, to control the difference between the counting function smoothed at small scales and the genuine counting function, we prove

$$|\rho_{h,T(h)} * N_h(1) - N_h(1)| \leq C_0/T(h),$$
$$N_h := \#\{\lambda_j \in \mathrm{Spec}(-h^2\Delta_g) : \lambda_j \leq 1\}. \tag{7.57}$$

Then, to control the distance between counting function smoothed at small scales and that smoothed at larger scales, we prove

$$|\rho_{h,T(h)} * N_h(1) - \rho_{h,t_0} * N_h(1)| \leq C_0/T(h), \tag{7.58}$$

for some $t_0 > 0$ independent of h.

Together, these estimates imply

$$|N_h(1) - \rho_{h,t_0} * N_h(1)| \leq C_0/T(h), \tag{7.59}$$

and one can use standard methods to prove Theorem 2.2.5 from (7.59).

In order to prove the estimate (7.57), we recall that

$$N_h(1) = \int_\Delta \mathbb{1}_{[0,1]}(-h^2\Delta_g)d\sigma_\Delta, \tag{7.60}$$

where, for an interval $I \subset \mathbb{R}$, $\mathbb{1}_I(-h^2\Delta_g)$ denotes the spectral projector for $-h^2\Delta_g$ onto the interval I, and

$$\Delta := \{(x,x) \in M \times M : x \in M\}.$$

For s small, we then think of the kernel of $\mathbb{1}_{[1-s,1]}(-h^2\Delta_g)$ as a quasimode for the operator

$$\mathbf{P} := P \otimes I, \qquad P := -h^2\Delta_g - 1,$$

acting on $M \times M$ and the integral over the diagonal as an integral of that quasimode over a submanifold. We are then in a position to use ideas from geodesic beams to prove (7.57).

To prove (7.58), we again use (7.60), but this time, we use the Fourier representation of $(\rho_{h,t_0} - \rho_{h,T}) * \mathbb{1}_{[0,\cdot]}(-h^2\Delta_g)$ together with Egorov's theorem to show that the contribution from non-periodic geodesics is small, and we use the same ideas used to prove (7.57) (thinking of the kernel as a quasimode and geodesic beam techniques) to show that the contribution from periodic geodesics is controlled by their volume.

We now proceed to the proof of Theorem 2.2.5. This proof is divided into several pieces. First, in Sect. 7.4.2 we give a Tauberian lemma which allows for comparison of $\rho_{h,T(h)} * N_h(1)$ and $N_h(1)$ under an assumption on the scale at which $N_h(t)$ behaves like a Lipschitz function together with the corresponding Lipschitz constant. In Sect. 7.4.4, we then prove the corresponding Lipschitz type estimates on N_h.

Next, in Sect. 7.4.5 we use the Fourier representations of $\rho_{h,T(h)} * N_h(1)$ and $\rho_{h,t_0} * N_h(1)$ to make the comparison (7.58). Finally, in Sect. 7.4.6 we combine the previous estimates to prove Theorem 2.2.5.

7.4.2 The Tauberian Lemma

Our first lemma will be used to compare smoothed and unsmoothed versions of the eigenvalue counting function. It roughly states that, if the function w_h behaves like a Lipschitz function down to scale σ_h^{-1}, then w_h and w_h smoothed at scale σ_h^{-1} are close to one another.

Lemma 7.4.1 *Let $\{K_j\}_{j=0}^{\infty} \subset \mathbb{R}_+$. Then, there exists $C > 0$ and for all $N_0 \in \mathbb{R}$, $N > 0$ there exists $C_N > 0$, such that the following holds. Let $\{\rho_h\}_{h>0} \subset \mathscr{S}(\mathbb{R})$ be a family of functions and $\{\sigma_h\}_{h>0} \subset \mathbb{R}_+$ such that for all $j \geq 1$ and $h > 0$,*

$$|\rho_h(s)| \leq \sigma_h K_j \langle \sigma_h s \rangle^{-j} \quad \text{for all } s \in \mathbb{R}.$$

Let $\{L_h\}_{h>0} \subset \mathbb{R}_+$, $\{B_h\}_{h>0} \subset \mathbb{R}_+$, $\{w_h : \mathbb{R} \to \mathbb{R}\}_{h>0}$, $I_h \subset [-K_0, K_0]$, $h_0 > 0$ and $\varepsilon_0 > 0$, be so that for all $0 < h < h_0$

- $|w_h(t - s) - w_h(t)| \leq L_h \langle \sigma_h s \rangle$ *for all $t \in I_h$ and $|s| \leq \varepsilon_0$,*
- $|w_h(s)| \leq B_h \langle s \rangle^{N_0}$ *for all $s \in \mathbb{R}$.*

Then, for all $0 < h < h_0$ and $t \in I_h$

$$\left| (\rho_h * w_h)(t) - w_h(t) \int_{\mathbb{R}} \rho_h(s) ds \right| \leq CL_h + C_N B_h \sigma_h^{-N} \varepsilon_0^{-N}.$$

Proof For all $0 < h < h_0$ and $t \in I_h$

$$\left| (\rho_h * w_h)(t) - w_h(t) \int_{\mathbb{R}} \rho_h(s) ds \right| = \left| \int_{\mathbb{R}} \rho_h(s) \big(w_h(t - s) - w_h(t) \big) ds \right|$$

$$\leq L_h \int_{|s| \leq \varepsilon_0} |\rho_h(s)| \langle \sigma_h s \rangle ds + B_h \int_{|s| \geq \varepsilon_0} |\rho_h(s)| \Big(\langle t - s \rangle^{N_0} + \langle t \rangle^{N_0} \Big) ds$$

$$\leq L_h \int_{|s| \leq \varepsilon_0} \sigma_h K_3 \langle \sigma_h s \rangle^{-2} ds + B_h \int_{|s| \geq \varepsilon_0} K_{N_0+2+N} \sigma_h \langle \sigma_h s \rangle^{-(N_0+2+N)} \Big(\langle t - s \rangle^{N_0} + \langle t \rangle^{N_0} \Big) ds.$$

The existence of C and C_N follows from integrability of each term and the boundedness of I_h. $\square$

7.4.3   Technical Estimates on the Spectral Projector

The following lemma is proved in [DG14, Lemma 3.11] and will be used below to show that the counting function is Lipschitz down to scale $T(h)^{-1}$. In what follows, we write $\|\cdot\|_{HS}$ for the Hilbert-Schmidt norm.

Lemma 7.4.2 *Let $V \subset S_\delta(T^*M; [0,1])$ be a bounded subset. Then, there are $C > 0$ and $h_0 > 0$, and for all $N > 0$ there exists $C_N > 0$, such that for all $\chi \in V$, $0 < h < h_0$, and $|s| \le 2h$,*

$$\|\mathbb{1}_{[1-s,1]}(-h^2\Delta_g)Op_h(\chi)\|^2_{HS} \le Ch^{1-n}\mu_{S^*M}(\operatorname{supp}\chi \cap S^*M) + C_N h^N, \quad (7.61)$$

$$h^{-2}\|P\mathbb{1}_{[1-s,1]}(-h^2\Delta_g)Op_h(\chi)\|^2_{HS} \le Ch^{1-n}\mu_{S^*M}(\operatorname{supp}\chi \cap S^*M) + C_N h^N. \quad (7.62)$$

In fact, only (7.61) is proved in [DG14, Lemma 3.11]. The inequality (7.62) follows from (7.61) after observing that

$$\|P\mathbb{1}_{[1-s,1]}(-h^2\Delta_g)Op_h(\chi)\|^2_{HS} \le \|P\mathbb{1}_{[1-s,1]}(-h^2\Delta_g)\|^2_{L^2 \to L^2}\|\mathbb{1}_{[1-s,1]}(-h^2\Delta_g)Op_h(\chi)\|^2_{HS}.$$

We will also need the following basic estimate for the trace of the spectral projector.

Lemma 7.4.3 *There are $N_0 > 0$, $C > 0$, $h_0 > 0$ such that for all $s \in \mathbb{R}$ and $0 < h < h_0$*

$$\left|\int_\Delta \mathbb{1}_{(-\infty,s]}(-h^2\Delta_g)d\sigma_\Delta\right| \le Ch^{-\frac{n}{2}}\langle s\rangle^{N_0}\|\mathbb{1}_{(-\infty,s]}(-h^2\Delta_g)\|_{L^2}.$$

Proof Observe that

$$\begin{aligned}
\left|\int \mathbb{1}_{(-\infty,s]}(-h^2\Delta_g)d\sigma_\Delta\right| &\\
= \operatorname{tr}\mathbb{1}_{(-\infty,s]}(-h^2\Delta_g) &= \operatorname{tr}(\mathbb{1}_{(-\infty,s]}(-h^2\Delta_g))^2 \\
= \|(\mathbb{1}_{(-\infty,s]}(-h^2\Delta_g))\|_{HS}{}^2 &= \|\mathbb{1}_{(-\infty,s]}(-h^2\Delta_g)\|_{HS}\|\mathbb{1}_{(-\infty,s]}(-h^2\Delta_g)\|_{L^2}
\end{aligned} \quad (7.63)$$

Next, for $N > \frac{n}{2}$,

$$\begin{aligned}
\|\mathbb{1}_{(-\infty,s]}(-h^2\Delta_g)\|^2_{HS} &= \operatorname{tr}(\mathbb{1}_{(-\infty,s]}(-h^2\Delta_g)) \\
&\le \|\mathbb{1}_{(-\infty,s]}(-h^2\Delta_g)\|_{\operatorname{tr}} \\
&\le \|(-h^2\Delta_g + 1)^{-N}\|_{\operatorname{tr}}\|(-h^2\Delta_g + 1)^N\mathbb{1}_{(-\infty,s]}(-h^2\Delta_g)\|_{L^2 \to L^2} \\
&\le Ch^{-n}(1 + |s|)^N.
\end{aligned} \quad (7.64)$$

In order to see the last line, we observe that by the functional calculus for the Laplacian (see e.g. [Zwo12, Chap. 14]), $(-h^2\Delta_g + 1)^{-N}$ is trace class and satisfies

$$\|(-h^2\Delta_g + 1)^{-N}\|_{\mathrm{tr}} \le Ch^{-n}.$$

The lemma now follows from (7.63) and (7.64). $\square$

7.4.4   Lipschitz Type Estimates for the Spectral Projector

In this section, we show that the spectral projector, or more precisely its trace, is Lipschitz to scale $T(h)^{-1}$.

We now introduce tubes over $N^*\Delta \cap (S^*M \times T^*M)$, where

$$N^*\Delta := \{(x, \xi, x, -\xi) \in T^*M \times T^*M \ : \ (x, \xi) \in T^*M\},$$

that we use when applying geodesic beam estimates. Although one can work with any tubes generated by the flow for $\mathbf{p}(x, \xi, y, \eta) := |\xi|^2_{g(x)} - 1$ (as in Sect. 6.2), because the operator $\mathbf{P}$ is the identity in one factor it is convenient in this section to work with tubes which factor; i.e. are of the form $Op_h(\chi_1) \otimes Op_h(\chi_2)$ for $\chi_i \in S_\delta(T^*M)$.

To describe these tubes, we work microlocally near a point $\rho_0 \in N^*\Delta \cap (S^*M \times T^*M)$. Let $\pi_R, \pi_L : T^*(M \times M) \to T^*M$ denote the projections to the right and left factor, and let $\tilde{\mathcal{Z}}_{\pi_L(\rho_0)} \subset T^*M$ be a transversal to the flow for $|\xi|^2_g$ containing $\pi_L(\rho_0)$. (Such a hypersurface exists since $d|\xi|^2_g(\rho) \ne 0$ on S^*M).) Define a transversal to the flow for $\mathbf{p}$ by

$$\mathcal{Z}_{\rho_0} := \tilde{\mathcal{Z}}_{\pi_L(\rho_0)} \times T^*M,$$

and let U be a neighborhood of ρ_0 in $N^*\Delta$ such that $U \cap (S^*M \times T^*M) \subset \mathcal{Z}_{\rho_0}$. We will use the metric $\tilde{d}$ on $T^*M \times T^*M$ defined by $\tilde{d}\big((\rho_L, \rho_R), (q_L, q_R)\big) := \max\big(d(\rho_L, q_L), d(\rho_R, q_R)\big)$, for $(\rho_L, \rho_R), (q_L, q_R) \in T^*M \times T^*M$. With this definition, for $\rho = (\rho_L, \rho_R) \in N^*\Delta \cap (S^*M \times T^*M)$, we put

$$\mathcal{T}_\rho := \Lambda_\rho^\tau(r) = \tilde{\Lambda}_{\rho_L}^\tau(r) \times B(\rho_R, r)$$

where $\Lambda_A^\tau(r)$ is defined by (6.13) (and the text below the equation) with φ_t the Hamiltonian flow for $\mathbf{p}$ and $\tilde{\mathcal{T}} = \tilde{\Lambda}_{\rho_L}^{\tau(r)}$ denotes a tube with respect to $|\xi|^2_g - 1$ and the hypersurface $\tilde{\mathcal{Z}}_{\pi_L(\rho_0)}$. In particular, when we use cutoffs with respect to a tube $\mathcal{T}$, we will always work with cutoffs of the form

$$\chi_{\mathcal{T}}(x, \xi, y, \eta) = \chi_{\tilde{\mathcal{T}}}(x, \xi)\chi_{\rho_R}(y, \eta), \qquad \operatorname{supp}\chi_{\rho_R} \subset B(\rho_R, r).$$

Throughout this section, we will assume that

$$[-h^2\Delta_g, Op_h(\chi_{\tilde{\mathcal{T}}})] \in h\Psi_\delta(M).$$

We can do this, for instance, by following the construction in Lemma 6.2.7. We will refer only to this tube in T^*M, leaving the other implicit and will think of the kernel of $A_1 \mathbb{1}_I(-h^2\Delta_g)A_2$ as that of $\mathbb{1}_I(-h^2\Delta_g)$ acted on by $A_1 \otimes A_2^t$.

Our next lemma shows that, when cutoff using an appropriate tube, the trace of the spectral projector is controlled by a Lipschitz type estimate times the volume of the tube in $S^*M \times T^*M$.

Lemma 7.4.4 *Suppose $0 < \varepsilon_0 < \frac{1}{2}$, $\mathfrak{D} > 0$, let τ_0, R_0, δ, $R(h)$ be as in Theorem 7.2.1, and $0 < \tau < \tau_0$. Let $\{\mathcal{T}_j\}_{j\in\mathcal{J}(h)}$ be a $(\mathfrak{D}, \tau, R(h))$ good covering of $(S^*M \times T^*M) \cap N^*\Delta$ and $\mathcal{V} \subset S_\delta(T^*M \times T^*M; [0,1])$ bounded. Then, there is $C_0 > 0$ such that for all $\{\chi_{\mathcal{T}_j}\}_{j\in\mathcal{J}(h)} \subset \mathcal{V}$ partitions for $\{\mathcal{T}_j\}_{j\in\mathcal{J}(h)}$, $j \in \mathcal{J}(h)$, and $|s| \leq \varepsilon_0$*

$$\left| \int_\Delta Op_h(\chi_{\mathcal{T}_j})\mathbb{1}_{[1-s,1]}(-h^2\Delta_g)d\sigma_\Delta \right| \leq C_0 h^{1-n} R(h)^{2n-1}\left\langle \frac{s}{h} \right\rangle.$$

Proof We first note that it suffices to prove the statement for $|s| \leq 2h$. Indeed, this is because we may decompose $\mathbb{1}_{[1-s,1]}(-h^2\Delta_g) = \sum_{k=0}^{k_0-1} \mathbb{1}_{[t_k,t_{k+1}]}(-h^2\Delta_g)$, with $|t_{k+1} - t_k| \leq h$ and apply the estimate for $\mathbb{1}_{[1-t_k/t_{k+1},1]}(-\tilde{h}^2\Delta_g)$ with $\tilde{h} = h/\sqrt{t_{k+1}}$ to obtain the result for $|s| \leq \varepsilon_0$.

Suppose $|s| \leq 2h$. Let, $j \in \mathcal{J}(h)$. We now apply (7.26) for some $(\mathfrak{D}, 2\tau, \tilde{R}(h))$ good cover of $(S^*M \times T^*M) \cap N^*\Delta$ with radius $\tilde{R}(h) \geq K R(h)$, $\{\tilde{\chi}_{\mathcal{T}_i}\}_{i\in\mathcal{I}}$, for some K chosen large enough such that, with e from (7.26), we have

$$Op_h(e)\mathbf{P}Op_h(\chi_{\mathcal{T}_j}) = O(h^\infty)_{\Psi-\infty}.$$

Then, with u the kernel of $\mathbb{1}_{[1-s,1]}(-h^2\Delta_g)$,

$$h^{\frac{n-1}{2}}\left| \int_\Delta Op_h(\chi_{\mathcal{T}_j})\mathbb{1}_{[1-s,1]}(-h^2\Delta_g)d\sigma_\Delta \right|$$
$$\leq C_0 R(h)^{\frac{2n-1}{2}} \sum_i \left(\|Op(\tilde{\chi}_{\mathcal{T}_i})Op_h(\chi_{\mathcal{T}_j})u\|_{L^2} + \frac{C}{h}\|Op(\tilde{\chi}_{\mathcal{T}_j})\mathbf{P}Op_h(\chi_{\mathcal{T}_j})u\|_{L^2} \right)$$
$$+ C_N h^N \|u\|_{L^2}$$

Now,

$$|\{i \in \mathcal{I} : \operatorname{supp}\tilde{\chi}_{\mathcal{T}_i} \cap \operatorname{supp}\chi_{\mathcal{T}_j} \neq \emptyset\}| \leq \mathfrak{D}.$$

Therefore,

$$h^{\frac{n-1}{2}}\left| \int_\Delta Op_h(\chi_{\mathcal{T}_j})\mathbb{1}_{[1-s,1]}(-h^2\Delta_g)d\sigma_\Delta \right|$$
$$\leq C_0 \mathfrak{D} R(h)^{\frac{2n-1}{2}} \left(\|Op_h(\chi_{\mathcal{T}_j})u\|_{L^2} + \frac{C}{h}\|\mathbf{P}Op_h(\chi_{\mathcal{T}_j})u\|_{L^2} \right) + C_N h^N \|u\|_{L^2}.$$

Now, for $\psi = \psi_1 \otimes \psi_2 \in S_\delta(T^*M \times M)$ with $\operatorname{supp}\psi \subset \mathcal{T}_j$, and $\psi \equiv 1$ on $\operatorname{supp}\chi_{\mathcal{T}_j}$,

$$\mathbf{P}Op_h(\chi_{\mathcal{T}_j})u = Op_h(\chi_{\mathcal{T}_j})Op_h(\psi)\mathbf{P}u + O(h)_{\Psi_\delta}Op_h(\psi)u + O(h^\infty)_{\Psi-\infty}u$$

since

$$[\mathbf{P}, Op_h(\chi_{T_j})] \in h\Psi_\delta(M \times M).$$

Therefore,

$$h^{\frac{n-1}{2}}\left|\int_\Delta Op_h(\chi_{T_j})\mathbb{1}_{[1-s,1]}(-h^2\Delta_g)d\sigma_\Delta\right|$$
$$\leq C_0 \mathfrak{D} R(h)^{\frac{2n-1}{2}}\left(\|Op_h(\psi_1)\mathbb{1}_{[1-s,1]}(-h^2\Delta_g)Op_h(\psi_2)\|_{HS} + \tfrac{C}{h}\|Op_h(\psi_1)\mathbf{P}\mathbb{1}_{[1-s,1]}(-h^2\Delta_g)Op_h(\psi_2)\|_{HS}\right)$$
$$\quad + C_N h^N \|\mathbb{1}_{[1-s,1]}(-h^2\Delta_g)\|_{HS}$$
$$\leq C h^{\frac{1-n}{2}} R(h)^{2n-1}.$$

In the last line we used Lemma 7.4.2 and the existence of $C > 0$ such that $\mu\big((\mathrm{supp}\,\psi_1) \cap S^*M\big) \leq C R(h)^{2n-1}$. This finishes the proof when $|s| \leq 2h$. $\qquad\square$

The next lemma shows that, under a dynamical assumption, the trace of the spectral projector is Lipschitz to scale $T(h)^{-1}$. The dynamical assumption is as follows. There exists a $(\mathfrak{D}, \tau, R(h))$ good cover of $(S^*M \times T^*M) \cap N^*\Delta$, $\{\mathcal{T}_j\}_{j\in\mathcal{J}}$, a partition of indices $\mathcal{J}(h) = \mathcal{G}_0(h) \sqcup \mathcal{G}_1(h)$, and a constant $C_{\mathrm{np}} > 0$ such that

$$\sum_{\ell=0}^{1}\sqrt{\frac{|\mathcal{G}_\ell(h)|}{T_\ell}} \leq \frac{C_{\mathrm{np}} R(h)^{\frac{1-2n}{2}}}{\sqrt{T(h)}}, \qquad \sum_{\ell=0}^{1}\sqrt{|\mathcal{G}_\ell(h)|T_\ell} \leq C_{\mathrm{np}} R(h)^{\frac{1-2n}{2}}\sqrt{T(h)}. \qquad (7.65)$$

It turns out that the existence of such a cover, with $T(h) = \mathbf{T}(R(h))$ is implied by the more natural dynamical assumption that M is $\mathbf{T}$ non-periodic in the sense of Definition 2.2.4. We will briefly discuss this in Chap. 8.

Lemma 7.4.5 *Let ε_0, τ_0, R_0, δ, $R(h)$ as in Theorem 7.2.1. Let $0 < \tau < \tau_0$, $0 < \gamma < \frac{1}{2} - \delta$, and $T(h) \leq \gamma T_e(h)$, and suppose that (7.65) holds. Then, there is $C_0 = C_0(n, P, C_{\mathrm{np}}, \varepsilon_0) > 0$ and $h_0 > 0$ such that for all $0 < h \leq h_0$, $|s| \leq 2h$,*

$$h^{n-1}\left|\int_\Delta \mathbb{1}_{[1-s,1]}(-h^2\Delta_g)d\sigma_\Delta\right|^2 \leq C_0 \frac{1}{T(h)}\left\langle\frac{T(h)s}{h}\right\rangle\|\mathbb{1}_{[1-s,1]}(-h^2\Delta_g)\|_{L^2}^2.$$

Proof Define

$$\tilde{T}_\ell^i(h) := \begin{cases} T_\ell^i(h)\left\langle\frac{T_i(h)s}{h}\right\rangle^{-1} & \left\langle\frac{T_i(h)s}{h}\right\rangle \leq T_\ell^i, \\ 1 & \text{else.} \end{cases}$$

Then, applying (7.27) there is $C_0 > 0$ so that

$$h^{\frac{n-1}{2}}\left|\int_{\Delta}\mathbb{1}_{[1-s,1]}(-h^2\Delta_g)d\sigma_\Delta\right|$$

$$\leq C_0 R(h)^{\frac{2n-1}{2}}\left(\sum_{\ell=0}\frac{(|\mathcal{G}_\ell(h)|)^{\frac{1}{2}}}{(\tau\tilde{T}_\ell)^{\frac{1}{2}}}\|u\|_{L^2}+\sum_\ell\frac{(|\mathcal{G}_\ell(h)|\tilde{T}_\ell)^{\frac{1}{2}}}{h}\|Pu\|_{L^2}\right)$$

$$+\,Ch^{-\frac{1}{2}}R(h)^{-1}\|Pu\|_{H_h^{\frac{n}{2}+1}}.$$

Finally, observe that

$$h^{-\frac{1}{2}}R(h)^{-1}\leq h^{-1}/T(h),$$

and

$$\|Pu\|_{H_h^{\frac{n}{2}+1}}\leq C\|Pu\|_{L^2},$$

from which the lemma follows. $\qquad\square$

7.4.5 The Fourier Representation to Compare Two Smoothed Counting Functions

In this section, we use the Fourier representation of a smooth counting function or, more precisely, a smoothed spectral projector, to estimate the difference between two smoothed spectral projectors; one smoothed at a small scale $(\sim hT(h)^{-1})$ and one at the larger scale h. As discussed above, this is possible through an application of Egorov's theorem and the method of geodesic beams.

Precisely, we let $\rho\in\mathcal{S}(\mathbb{R})$ with $\hat{\rho}(0)\equiv 1$ on $[-1,1]$ and $\mathrm{supp}\,\hat{\rho}\subset[-2,2]$. For $T>0$ and $t_0>0$ we define

$$\rho_{h,T}(t):=\tfrac{T}{h}\,\rho\!\left(\tfrac{T}{h}t\right),\tag{7.66}$$

and

$$f_{T,h}(\lambda):=f_T(h^{-1}\lambda),\qquad f_T(\lambda):=\frac{1}{i}\int_{\mathbb{R}}\frac{1}{\tau}\hat{\rho}\!\left(\tfrac{\tau}{T}\right)\!\left(1-\hat{\rho}\!\left(\tfrac{\tau}{t_0}\right)\right)e^{-i\tau\lambda}d\tau,\tag{7.67}$$

where T is a positive constants with $t_0<T$, and ρ is as in (7.66).

We begin by providing some elementary estimates on f_T.

Lemma 7.4.6 *We note that for all $N>0$*

$$|f_T(\lambda)|\leq C_N\langle\lambda\rangle^{-N},\qquad \mathrm{supp}\,\hat{\rho}\!\left(\tfrac{\tau}{T}\right)\!\left(1-\hat{\rho}\!\left(\tfrac{\tau}{t_0}\right)\right)\subset\{\tau\in\mathbb{R}:|\tau|\in[t_0,2T]\}.\tag{7.68}$$

Proof The support properties of $\mathrm{supp}\,\hat{\rho}\!\left(\tfrac{\tau}{T}\right)\!\left(1-\hat{\rho}\!\left(\tfrac{\tau}{t_0}\right)\right)$ follow directly from the properties of ρ. For the estimate, first observe that

$$|\lambda^N f_T(\lambda)| = \left| \int_{\mathbb{R}} \frac{1}{\tau} \hat{\rho}(\tfrac{\tau}{T})(1 - \hat{\rho}(\tfrac{\tau}{t_0}))\lambda^N e^{-i\tau\lambda}d\tau \right|$$

$$= \left| \int_{\mathbb{R}} D_\tau^N\left(\frac{1}{\tau}\hat{\rho}(\tfrac{\tau}{T})(1 - \hat{\rho}(\tfrac{\tau}{t_0}))\right)e^{-i\tau\lambda}d\tau \right|$$

$$\leq C_N.$$

Therefore, we only need to prove that $|f_T(\rho)| \leq C$. For this, first note that

$$|f_T(\lambda)| \leq \int_{t_0 \leq |\tau| \leq 2t_0} \left| \frac{1}{\tau}\hat{\rho}(\tfrac{\tau}{T})(1 - \hat{\rho}(\tfrac{\tau}{t_0})) \right| d\tau + \int_{T \leq |\tau| \leq 2T} \left| \frac{1}{\tau}\hat{\rho}(\tfrac{\tau}{T})(1 - \hat{\rho}(\tfrac{\tau}{t_0})) \right| d\tau$$

$$+ \left| \int_{2|t_0| \leq \tau \leq |T|} \frac{1}{\tau}e^{-i\tau\lambda}d\tau \right|$$

$$\leq C + \left| \int_{2t_0}^T \frac{1}{\tau}e^{-i\tau\lambda}d\tau + \int_{2t_0}^T \frac{1}{-\tau}e^{i\tau\lambda}d\tau \right|$$

$$\leq C + 2\left| \int_{2t_0}^T \frac{\sin(\tau)}{\tau}d\tau \right| \leq C. \qquad \Box$$

In our next lemma, we rewrite the difference between two smoothed spectral projectors in terms of the function $f_{T,h}$. Here and below, for $s \in \mathbb{R}$, we write

$$\Pi_h(s) := \mathbb{1}_{(-\infty,s]}(-h^2\Delta_g).$$

Lemma 7.4.7 *For all $N > 0$, and $P := -h^2\Delta_g - 1$,*

$$(\rho_{h,\tilde{T}} - \rho_{h,t_0}) * \Pi_h(1) = f_{\tilde{T},h}(P) + O(h^N)_{H_{\mathrm{scl}}^{-N} \to H_{\mathrm{scl}}^N}.$$

Proof In this proof, we write $P_E := -h^2\Delta_g - E$. First, we prove that for $E_1, E_2 \in \mathbb{R}$, then

$$\int_{E_1}^{E_2} (\rho_{h,\tilde{T}(h)} - \rho_{h,t_0}) * \partial_s \Pi_h(s)ds = f_{\tilde{T}(h),h}(P_{E_2}) - f_{\tilde{T}(h),h}(P_{E_1}). \qquad (7.69)$$

To ease notation write $\tilde{T}$ for $\tilde{T}(h)$. To prove (7.69) we write

$$\int_{E_1}^{E_2} (\rho_{h,\tilde{T}} - \rho_{h,t_0}) * \partial_s \Pi_h(s)ds = \int_{E_1}^{E_2} \int_{\mathbb{R}} \hat{\rho}(\tfrac{w}{\sigma_{h,\tilde{T}}})[1 - \hat{\rho}(\tfrac{w}{\sigma_{h,t_0}})]e^{-iw(-h^2\Delta_g - s)}dwds,$$

where we use $\hat{\rho}(\tfrac{w}{\sigma_{h,t_0}}) = \hat{\rho}(\tfrac{w}{\sigma_{h,\tilde{T}}})\hat{\rho}(\tfrac{w}{\sigma_{h,t_0}})$. Putting $\tau := hw$, (7.69) follows.

Next, let $N > 0$. By (7.69) it suffices to find $E_1 \in \mathbb{R}$ such that

$$\left\| f_{\tilde{T},h}(P_{E_1}) \right\|_{H_{\mathrm{scl}}^{-N} \to H_{\mathrm{scl}}^N} \leq C_N h^{2N}, \qquad \|\rho_{h,t} * \Pi_h(E_1)\|_{H_{\mathrm{scl}}^{-N} \to H_{\mathrm{scl}}^N} = O(h^N). \qquad (7.70)$$

To prove the first claim in (7.70), note that by (7.68) for all $N > 0$ there is $C_N > 0$ such that

$$\left\| P_{E_1}^{\ N} f_{\widetilde{T},h}(P_{E_1}) P_{E_1}^{\ N} \right\|_{L^2 \to L^2} \leq C_N h^{2N}.$$

Next, observe that $\sigma(-h^2 \Delta_g) = |\xi|_g^2 \geq 0$ for all $(x, \xi) \in T^*M$. In particular, for $E_1 \leq -2$, P_{E_1} is elliptic and we have $P_{E_1}^{-1} : H_{\mathrm{scl}}^s(M) \to H_{\mathrm{scl}}^{s+k}(M) = O_s(1)$ for all $s \in \mathbb{R}$. Then, for $E_1 \leq -2$ the first claim in (7.70) follows.

Next, by the sharp Gårding inequality, there is $C > 0$ such that $\Pi_h(s) \equiv 0$ for $s \leq -Ch$. Thus, for $E_1 \leq -3$ and all $N, M \geq 0$ there is $C_{M,N} > 0$ such that

$$\|(\rho_{h,t} * \Pi_h)(E_1)\|_{H_{\mathrm{scl}}^{-N} \to H_{\mathrm{scl}}^N} \leq \int_{\mathbb{R}} \tfrac{t}{h} \rho\big(\tfrac{t}{h}s\big) \|\Pi_h(E_1 - s)\|_{H_{\mathrm{scl}}^{-N} \to H_{\mathrm{scl}}^N} ds$$

$$\leq C_{M,N} \int_{s \leq -1} \tfrac{t}{h}\big\langle \tfrac{t}{h}s\big\rangle^{-M} \langle s \rangle^{2N/k}.$$

The claim follows after choosing M large enough. $\qquad\square$

Now that we have written Fourier representation of the difference between two spectral projectors, we can apply geodesic beam methods to estimate the contribution from an individual tube in phase space. This is the content of the next lemma.

Lemma 7.4.8 *Let* $\mathcal{I} \subset \mathcal{J}(h)$. *Then, there exist* $C_0 > 0$ *and* $h_0 > 0$ *such that for all* $0 < h < h_0$

$$\left| \int_\Delta \sum_{j \in \mathcal{I}} Op_h(\chi_{T_j}) f_{\widetilde{T},h}(P) d\sigma_\Delta \right| \leq C_0 h^{1-n} R(h)^{2n-1} |\mathcal{I}|.$$

Proof We first note that $f_{\widetilde{T}(h),h}(P) = \varrho_h * \partial_s \mathbb{1}_{(-\infty, \cdot]}(P)(1)$, where $\varrho_h(s) := f_{\widetilde{T}(h),h}(-s)$. Then, since $\widehat{f_{\widetilde{T}(h)}}(0) = 0$, we have $\int_{\mathbb{R}} \partial_s \varrho_h(s) ds = 0$. In particular, by the estimates (7.68), Lemma 7.4.1 applies with $\sigma_h = h^{-1}$. Note that by Lemma 7.4.4, for $|s| \leq 1$,

$$\left| \int_\Delta Op_h(\chi_{T_j})(\mathbb{1}_{(-\infty,1]}(-h^2 \Delta_g) - \mathbb{1}_{(-\infty,1-s]}(-h^2 \Delta_g)) d\sigma_\Delta \right| \leq C h^{1-n} R(h)^{2n-1}\big\langle \tfrac{s}{h}\big\rangle.$$
$$\tag{7.71}$$

Also, by Lemma 7.4.3, there exists N_0 such that for $s \in \mathbb{R}$,

$$\left| \int_\Delta Op_h(\chi_{T_j}) \mathbb{1}_{(-\infty,s]}(-h^2 \Delta_g) d\sigma_\Delta \right| \leq C h^{-n} \langle s \rangle^{N_0}. \tag{7.72}$$

The proof follows from Lemma 7.4.1 using (7.71) and (7.72), and by summing in $j \in \mathcal{I}$. $\qquad\square$

Next, again using the Fourier representation from Lemma 7.4.7, we use Egorov's theorem to show that tubes which do not return to themselves under the flow do not contribute significantly.

Lemma 7.4.9 *Suppose $\mathcal{T}_j$ is a tube such that $\tilde{\mathcal{T}}_j$, its corresponding tube in T^*M, satisfies $\varphi_t(\tilde{\mathcal{T}}_j) \cap \tilde{\mathcal{T}}_j = \emptyset$ for $|t| \in [t_0, T(h)]$. Then for all $N > 0$ there is $C_N > 0$ such that*

$$\left| \int_\Delta Op_h(\chi_{\mathcal{T}_j}) f_{\tilde{T},h}(P) d\sigma_\Delta \right| \leq C_N h^N.$$

Proof Note that the assumption on $\tilde{\mathcal{T}}_j$ implies $\exp(t H_{\mathbf{p}})(\tilde{\mathcal{T}}_j) \cap \Lambda^\tau_{N*\Delta}(R(h)) = \emptyset$ for $|t| \in [t_0, T(h)]$.

By Egorov's Theorem 3.4.1, for all $N > 0$ there exist $h_0 > 0$ and $C_N > 0$ such that for all $0 < h < h_0$,

$$\left\| Op_h(\chi_{\mathcal{T}^1_j}) e^{-it \frac{P}{h}} Op_h(\chi_{\mathcal{T}^2_\ell}) \right\|_{H^{-N}_{\mathrm{scl}}(M) \to H^N_{\mathrm{scl}}(M)} \leq C_N h^N, \qquad |t| \in [t_0 + \tau, T(h) - \tau].$$

The claim follows from the definition (7.67) together with the facts that by (7.68) the support of its integrand has $\tau \in [t_0, 2\tilde{T}(h)]$, and $\tilde{T}(h) = \frac{1}{2} T(h)$. $\qquad\qquad\square$

7.4.6 Proof of Theorem 2.2.5

We now prove the main theorem of this section. To do this, we claim that

$$h^{n-1} \left| \int_\Delta \Big(\Pi_h(1) - (\rho_{h,t_0} * \Pi_h)(1) \Big) d\sigma_\Delta \right| \leq C_0 / T(h). \tag{7.73}$$

We start by showing under the same assumptions that

$$h^{n-1} \left| \int_\Delta \Big((\rho_{h,T(h)} * \Pi_h(1) - \Pi_h(1) \Big) d\sigma_\Delta \right| \leq C_0 / T(h), \tag{7.74}$$

$$h^{n-1} \left| \int_\Delta \Big((\rho_{h,T(h)} * \Pi_h)(1) - (\rho_{h,t_0} * \Pi_h(1) \Big) d\sigma_\Delta \right| \leq C_0 / T(h). \tag{7.75}$$

for some t_0 independent of h. We will derive Theorem 2.2.5 from (7.73).

7.4.6.1 Proof of (7.74)

Then, for $|s| \leq 2h$, by Lemma 7.4.5

$$h^{n-1} \operatorname{tr} \left(\mathbb{1}_{[1-s,1]}(-h^2 \Delta_g) \right)^2 \left(\frac{1}{T(h)} \Big\langle \frac{T(h)s}{h} \Big\rangle \right)^{-1}$$

$$\leq C_0 \| \mathbb{1}_{[1-s,1]}(-h^2 \Delta_g) \|^2_{L^2} = C_0 \operatorname{tr} \mathbb{1}_{[1-s,1]}(-h^2 \Delta_g).$$

Hence, for $|s| \leq 2h$ we have

$$0 \leq h^{n-1} \operatorname{tr} \mathbb{1}_{[1-s,1]}(-h^2 \Delta_g) \leq \frac{C_0}{T(h)}\left\langle \frac{T(h)s}{h} \right\rangle.$$

Next, for $|s| \leq \varepsilon_0$, splitting $\mathbb{1}_{[1-s,1]}(-h^2 \Delta_g) = \sum_{k=0}^{k_0-1} \mathbb{1}_{[t_k, t_{k+1}]}(-h^2 \Delta_g)$ with

$$t_0 = \min(t-s,t), \quad t_k = t_{k-1} + h, \ 1 \leq k < k_0, \quad t_{k_0} = \max(t-s,t), \quad k_0 := \left\lceil \tfrac{|s|}{h} \right\rceil,$$

and replace h by $\tilde{h}_k := h/\sqrt{t_k}$, so that $t_{k+1}/t_k \leq 2\tilde{h}_k$ and we may apply our results with $|s| \leq 2\tilde{h}_k$. We have by Lemmas 7.4.5 and 7.4.3 that there exists $N_0 > 0$ such that

$$h^{n-1}\left| \int_{\boldsymbol{\Delta}} \mathbb{1}_{[1-s,1]}(-h^2 \Delta_g) d\sigma_{\boldsymbol{\Delta}} \right| \leq C_0 \frac{1}{T(h)}\left\langle \frac{T(h)s}{h} \right\rangle, \tag{7.76}$$

$$h^{\frac{n}{2}}\left| \int_{\boldsymbol{\Delta}} \Pi_h(s) d\sigma_{\boldsymbol{\Delta}} \right| \leq C \langle s \rangle^{N_0} \|\Pi_h(s)\|_{L^2} \leq C h^{-\frac{n}{2}}(1 + |s|^{2N_0}). \tag{7.77}$$

In particular, combining (7.76) and (7.77) together with Lemma 7.4.1 implies (7.74) holds.

7.4.6.2 Proof of (7.75)

Using Lemma 7.4.7, the proof of (7.75) amounts to understanding

$$\left((\rho_{h,\tilde{T}(h)} - \rho_{h,t_0}) * \Pi_h(1) = f_{\tilde{T}(h),h}(P) + O(h^\infty) \right)_{H_{\mathrm{scl}}^{-N} \to H_{\mathrm{scl}}^N},$$

where $f_{T,h}$ is given by (7.67), and $\tilde{T}(h) = \frac{T(h)}{2}$. In particular, we consider $\operatorname{tr} f_{\tilde{T},h}(P)$. For this, we let $\{\mathcal{T}_j\}_{j \in \mathcal{J}(h)}$ be a $(\mathfrak{D}, \tau, R(h))$-good covering of $N^*\boldsymbol{\Delta} \cap (S^*M \times T^*M)$ and $\mathcal{V} \subset S_\delta(T^*M \times M; [0,1])$ a bounded subset. Let $\{\chi_{\mathcal{T}_j}\}_{j \in \mathcal{J}(h)} \subset \mathcal{V}$ be a good partition associated to $\{\mathcal{T}_j\}_{j \in \mathcal{J}_E(h)}$.

Since (7.65) holds, there is a splitting $\mathcal{J}(h) = \mathcal{B}(h) \cup \mathcal{G}(h)$ such that $\varphi_t(\tilde{\mathcal{T}}_j) \cap \tilde{\mathcal{T}}_j = \emptyset$ for $|t| \in [t_0, T(h)]$ for $j \in \mathcal{G}(h)$, and $|\mathcal{B}(h)| R(h)^{2n-1} \leq T^{-1}(h)$. We write, using $\Lambda^\tau_{S^*M \times T^*M \cap N^*\boldsymbol{\Delta}}(R(h)/2) \subset \bigcup_{j \in \mathcal{J}_h} \mathcal{T}_j$,

$$\int_{\boldsymbol{\Delta}} f_{\tilde{T},h}(P) d\sigma_{\boldsymbol{\Delta}} = \sum_{j \in \mathcal{G}(h) \cup \mathcal{B}(h)} \int_{\boldsymbol{\Delta}} Op_h(\chi_{\mathcal{T}_j}) f_{\tilde{T},h}(P) d\sigma_{\boldsymbol{\Delta}} + O(h^\infty).$$

Applying Lemma 7.4.9 to the sum over $\mathcal{G}(h)$ and Lemma 7.4.8 to the sum over $\mathcal{B}(h)$, we have

$$\left| \int_{\boldsymbol{\Delta}} f_{\tilde{T},h}(P) d\sigma_{\boldsymbol{\Delta}} \right| \leq C h^{1-n} |\mathcal{B}(h)| R(h)^{2n-1} + O(h^\infty) \leq C h^{1-n}/T(h).$$

In particular (7.75) holds.

7.4.6.3 Proof of Theorem 2.2.5

We assume (7.65) holds so we may apply (7.73). Let $0 < \delta < \frac{1}{2}$.

Let $R(h) \geq h^{\delta}$, and $T(h) = \mathbf{T}(R(h))$. Then, using (7.73) we obtain

$$\left| \int_{\Delta} \Big(\Pi_h(1) - \rho_{h,t_0} * \Pi_h(1) \Big) d\sigma_{\Delta} \right| \leq C_0 h^{1-n} / T(h).$$

In particular,

$$\left| \int \Pi_h(1, x, x) \, dv_g(x) - (2\pi h)^{-n} \, \mathrm{vol}_{\mathbb{R}^n}(B^n) \, \mathrm{vol}_g(M) \right| \leq C h^{1-n} T(h)^{-1}. \tag{7.78}$$

Now observe that

$$\int \Pi_h(1, x, x) \, dv_g(x) = \mathrm{tr}\left(\mathbb{1}_{(-\infty,1]}(-h^2 \Delta_g) \right) = \mathrm{tr}\left(\mathbb{1}_{(-\infty,h^{-2}]}(-\Delta_g) \right)$$

$$= \#\{ \lambda_j^2 \in \mathrm{Spec}(-\Delta_g) \,:\, \lambda_j^2 \leq h^{-2} \} = N(h^{-1}),$$

and hence, setting $\lambda = h^{-1}$, (7.78) reads

$$\left| N(\lambda) - (2\pi)^{-n} \lambda^n \, \mathrm{vol}_{\mathbb{R}^n}(B^n) \, \mathrm{vol}_g(M) \right| = |E(\lambda)| \leq C \frac{\lambda^{n-1}}{T(\lambda)}.$$

This completes the proof of Theorem 2.2.5. $\square$

7.4.7   Pointwise Weyl Asymptotics

The proof of Theorem 2.2.7 and many more general asymptotic formulae appear in [CG20b]. Their proofs follow the same structure as that of Theorem 2.2.5. The main difference in the approach is that we need not apply the geodesic beam method in $M \times M$ with the operator $\mathbf{P}$. Instead, we can apply them directly to estimate, for some $x \in M$, the quantity

$$|\mathbb{1}_{[1-s,t]}(-h^2 \Delta_g) v|(x),$$

in terms of the L^2 norm of v. This type of estimate can then be used to prove the analog of Lemma 7.4.5. In particular, that

$$h^{n-1} \left| \mathbb{1}_{[1-s,t]}(-h^2 \Delta_g)(x, y) \right| \leq C_0 \frac{1}{T(h)} \left(\frac{T(h)s}{h} \right),$$

under the assumptions of Theorem 2.2.7.

As in the proof of Theorem 2.2.5, it then remains to estimate

$$f_{\tilde{T},h}(P)(x,y).$$

This can again be done through a combination of Egorov's thoerem and the geodesic beam estimate on

$$\left| Op_h(\chi_{\mathcal{T}_j}) f_{\tilde{T},h}(P) Op_h(\chi_{\mathcal{T}_k}) \right|(x,y),$$

for $\mathcal{T}_j$ and $\mathcal{T}_k$ tubes in a $(\mathfrak{D},\tau,R(h))$ good cover of $S_x^* M$ and $S_y^* M$ respectively. One obtains an estimate of the form

$$\left| f_{\tilde{T},h}(P)(x,y) \right| \le Ch^{1-n}|\mathcal{B}(h)|R(h)^{n-1} + O(h^{\infty}) \le Ch^{1-n}/T(h),$$

where $\mathcal{B}(h)$ is the collection of tubes whose flow-outs pass over $S_y^* M$ for times in $[t_0, T(h)]$.

This concludes the proof of Theorem 2.2.5. $\qquad\square$

Dynamical Ideas 8

In this chapter we do not attempt to give complete proofs of any dynamical statements. Instead, our goal is only to indicate the ideas that go into the decomposition of good covers into collections which provide useful estimates in Theorems 7.1.5 and 7.2.2. For the details of the proofs, we refer the readers to [CG20b] and [CG19a]. Although the ideas remain the same, because one looks for quantitative information about sets undergoing the geodesic flow for large times, the precise proofs become substantially more complicated.

The chapter is divided into three sections. We first discuss how to pass from assumptions on the volume of periodic or looping trajectories to decompositions of tubes. We then illustrate the two main ideas used to construct covers: transversality and contraction.

8.1 Designing Coverings to Apply Volume Assumptions on Non-looping Directions

In the Weyl Law results, such as Theorems 2.2.7 and 2.2.5, as well as the L^p results, Theorems 2.2.11 and 2.2.12 we use hypotheses of the form

$$\limsup_{R \to 0^+} \mu_{S^*M}\Big(B_{S^*M}\big(\mathcal{P}_U^R(t_0, \mathbf{T}(R)), R\big) \Big) \mathbf{T}(R) \le C_{\mathrm{np}}. \tag{8.1}$$

and

$$\limsup_{R \to 0^+} \Big(\mu_{S_{\tilde{x}}^*M}\Big(B_{S_{\tilde{x}}^*M}(\mathcal{L}_{x,y}^R(t_0, \mathbf{T}(R)), R)\Big) \mu_{S_{\tilde{y}}^*M}\Big(B_{S_{\tilde{y}}^*M}(\mathcal{L}_{y,x}^R(t_0, \mathbf{T}(R)), R)\Big) \mathbf{T}(R)^2 \Big) \le C_{\mathrm{nl}}. \tag{8.2}$$

However, we have proved our estimates in Chap. 7 in terms of decompositions of covers by tubes into subsets of tubes with the non-self looping property. Passing from this type of cover assumption to a volume assumption like (8.1) and (8.2) is done as follows.

We consider the case of (8.2) with $x = y$ for concreteness. Let $\{\mathcal{T}_j\}_{j \in \mathcal{J}}$ be a $(\mathfrak{D}, \tau, R)$ good cover of $S_x^* M$. Then, we define the 'bad' set of tubes which will have no good non-looping property as

$$\mathcal{B} := \{j \in \mathcal{J} : \mathcal{T}_j \cap B_{S_x^* M}(\mathcal{L}_{x,x}^{4R}(t_0, T), 2R) \neq \emptyset\}.$$

The definition of a good cover then implies that

$$|\mathcal{B}| \leq C R^{1-n} \mu_{S_x^* M}\left(B_{S_x^* M}(\mathcal{L}_{x,x}^{4R}(t_0, T), 4R)\right),$$

and it remains only to check that, with $\mathcal{G} := \mathcal{J} \setminus \mathcal{B}$,

$$\bigcup_{j \in \mathcal{G}} \mathcal{T}_j \quad \text{is } [t_0 + 2(\tau + R), T - 2(\tau + R)] \text{ non-self looping.}$$

Then, for $j \in \mathcal{G}$, $\mathcal{T}_j(R) = \Lambda_{\rho_0}^\tau(R)$ for some $\rho_0 \in S_x^* M$ and $d(\rho_0, \mathcal{L}_{x,x}^{4R}(t_0, T)) > 3R$. In particular, since $\bigcup_{t_0 \leq |t| \leq T} \varphi_t(B(\rho_0, 4R)) \cap B(S_x^* M, 4R) = \emptyset$ and $\mathcal{T}_j \subset \bigcup_{|t| \leq \tau + R} \varphi_t(B(\rho_0, 3R))$, this yields

$$\bigcup_{t_0 + \tau + r \leq |t| \leq T - (\tau + r)} \varphi_t(\mathcal{T}_j) \cap B(S_x^* M, 4R) = \emptyset. \tag{8.3}$$

On the other hand, since for all $k \in \mathcal{J}$, we have $\mathcal{T}_k \cap \mathcal{S} \subset B(S_x^* M, 3R)$,

$$\mathcal{T}_k \subset \bigcup_{|t| \leq \tau + R} \varphi_t(B(S_x^* M, 3R)) \tag{8.4}$$

In particular, combining (8.3) and (8.4) we have

$$\bigcup_{t_0 + 2(\tau + R) \leq |t| \leq T - 2(\tau + R)} \varphi_t(\mathcal{T}_j) \cap B(S_x^* M, 4R) = \emptyset.$$

By this construction, if (8.2) holds (with $x = y$), then Theorem 7.1.5 yields that for an eigenfunction, u and a sublogarithmic rate function $\mathbf{T}(R)$,

$$|u(x)| \leq C h^{\frac{1-n}{2}} R(h)^{\frac{n-1}{2}} \left(|\mathcal{B}|^{1/2} + \frac{|\mathcal{G}|^{\frac{1}{2}}}{(\mathbf{T}(4R(h)))^{1/2}}\right) \|u\|_{L^2}$$

$$\leq C h^{\frac{1-n}{2}} \left(\sqrt{\mu_{S_x^* M}\left(B_{S_x^* M}(\mathcal{L}_{x,x}^{4R}(t_0, \mathbf{T}(4R(h))), 4(R(h)))\right)} + \frac{1}{\sqrt{\mathbf{T}(4R(h))}}\right) \|u\|_{L^2}$$

$$\leq \frac{C h^{\frac{1-n}{2}}}{\sqrt{\mathbf{T}(4R(h))}} \|u\|_{L^2}.$$

In particular, using that $\mathbf{T}$ is sublogarithmic, we have $\mathbf{T}(h^{-1}) \leq C \mathbf{T}(4R(h))$ and this yields Theorem 2.2.12.

8.2 Transversality

The first main tool used to decompose $(\mathfrak{D}, \tau, R)$ covers of SN^*H into non-self looping collections is, for a point $\rho \in SN^*H$, the transversality of $d\varphi_t|_\rho(T\,SN^*H)$ and $T_{\varphi_t(\rho)}SN^*H$. Indeed, if these two sets are transverse then, locally near ρ, there is a direction $v \in T_\rho SN^*H$, such that for $\Gamma : (-\varepsilon, \varepsilon) \to SN^*H$ with $\dot{\Gamma}(0) = v$,

$$\varphi_{t'}(\Gamma(s)) \notin SN^*H, \qquad 0 < |s| < \varepsilon_1, \quad |t' - t| < c.$$

Then, under such assumptions, the set of $\rho \in SN^*H$ for which $\varphi_{t'}(\rho) \in SN^*H$ for some t' near t is contained in a smooth, codimension 1 submanifold of SN^*H and hence can be covered by a small number of balls (which are then extended to tubes in T^*M).

In order to construct an effective cover for use in our estimates one, of course, needs quantitative estimates on the degree of transversality between $d\varphi_t|_\rho(T\,SN^*H)$ and $T_{\varphi_t(\rho)}SN^*H$. This allows both for control on the size of ε_1 in terms of t as well as the size of the ball around ρ in SN^*H within which this transversality is stable and hence, within which, the set of points returning to SN^*H is contained in a codimension 1 submanifold. For the precise version of this argument see [CG19a, Sect. 2].

We now sketch how this type of transversality is achieved in the case that $H = \{x\}$ for some $x \in M$ such that

$$d\left(x, C_x^{r_t, t}\right) \geq r_t, \qquad \text{for } t \geq t_0, \qquad r_t := 1/ae^{-at} \tag{8.5}$$

where $C_x^{r_t, t}$ is defined in (2.8).

8.2.1 Sketch of the Proof of Theorem 2.2.1

First observe that x being conjugate to itself along a geodesic of length t is precisely the statement that there are $\rho \in S_x^* M$ and $v \in T_\rho S_x^* M$ such that $d\varphi_{t/2}|_\rho v \in T_{\varphi_{t/2}(\rho)} S_x^* M$. (The reason for the scaling $t/2$ is that we have defined φ_t as the Hamiltonian flow for $|\xi|_g^2$ rather than $|\xi|_g$.) So, in fact, if x is not maximally self conjugate, in the sense that *all* perpendicular Jacobi fields along some geodesic γ vanish when $\gamma(t) = x$, then for every $\rho \in S_x^* M$ and t such that $\varphi_{t/2}(\rho) \in S_x^* M$, there is $v \in T_\rho S_x^* M$ such that $d\varphi_{t/2}|_\rho v \notin T_{\varphi_{t/2}(\rho)} S_x^* M$. In particular, the required transversality holds. What remains, then, is to get a quantitative estimate on the angle between $d\varphi_{t/2}v$ and $T S_x^* M$. It is the need for this quantitative estimate which requires us to assume that at least one Jacobi field does not vanish in a small (t dependent) neighborhood around x. The structure of the Jacobi equation (or more precisely an associated Ricatti equation) allows one to convert a lower bound on the size of the ball in which one has at least one non-vanishing Jacobi field into an estimate on how large that Jacobi field must be at x and hence into the requisite lower bound on the angle. The careful

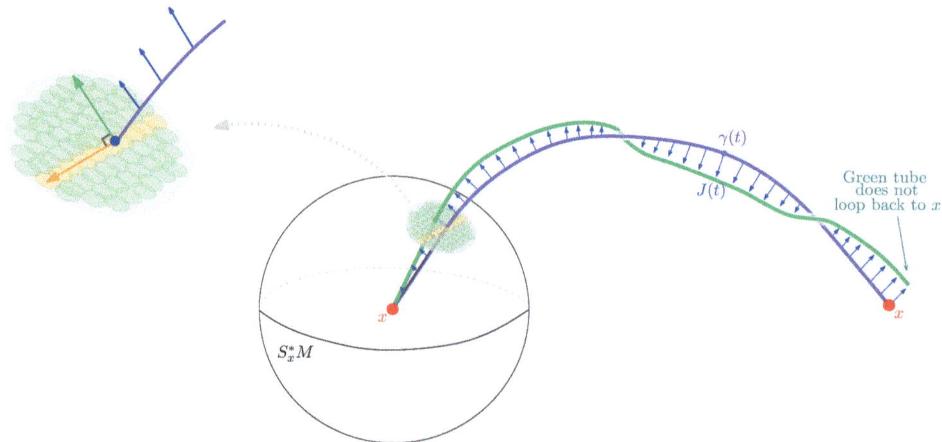

Fig. 8.1 In the figure, $H = \{x\}$ and $SN^*H = S_x^*M$. We consider a geodesic γ (purple curve) looping through $x = \gamma(0) = \gamma(t)$, and we assume that it can be perturbed in the direction of a vector $v \in S_x^*M$ (green arrow) yielding a new geodesic (green curve) along which there is a Jacobi field that does not vanish at time t. The point $\rho \in S_x^*M$ in the discussion above corresponds to the pair $(x, \gamma'(0))$. The picture illustrates how, at most, one will have a codimension one set of 'bad' directions, the one orthogonal to v (orange arrow), in which γ could possibly be perturbed to yield a geodesic that returns to x in time close to t. Indeed, only Jacobi fields whose derivative at time 0 lies in the orange plane could possibly vanish at x. With this structure in hand, we may cover the orthogonal direction to v using 'bad' tubes (orange balls) and then cover the remaining directions with 'good' tubes (green balls)

proof of these estimates can be found in [CG19a, Sect. 4]. (See Fig. 8.1 for a schematic illustration of this construction in the case.)

At the end of this construction we see that, under the condition (8.5), given a $(\mathfrak{D}, \tau, R)$ cover of S_x^*M, $\{\mathcal{T}_j\}_{j \in \mathcal{J}}$ and a point $\rho \in S_x^*M$ such that $d(\varphi_t(\rho), S_x^*M) < e^{-Ct}$, if $R \ll e^{-Ct}$ there are at most $\sim R^{2-n}e^{-C(n-2)t}$ tubes $\mathcal{T}_j$ which intersect $B(\rho, e^{-Ct})$ such that

$$\bigcup_{|s-t|<c} \varphi_s(\mathcal{T}_j) \cap \bigcup_{\ell \in \mathcal{J}} \mathcal{T}_\ell \neq \emptyset. \tag{8.6}$$

Therefore, covering S_x^*M by balls of radius e^{-Ct}, there are there are at most $R^{2-n}e^{Ct}$ tubes such that (8.6) holds. Running this argument for each $t \in \{t_0, t_0 + c, t_0 + 2c, \ldots, T\}$, we see that, with

$$\mathcal{B} := \left\{\ell : \bigcup_{s=t_0}^{T} \varphi_s(\mathcal{T}_j) \cap \bigcup_{\ell \in \mathcal{J}} \mathcal{T}_\ell \neq \emptyset\right\}, \qquad \mathcal{G} := \mathcal{J} \setminus \mathcal{B} \tag{8.7}$$

we have

$$|\mathcal{B}| \leq R^{2-n}e^{CT}T \quad \text{and} \quad \mathcal{G} \text{ is} [t_0, T] \text{non-self looping.}$$

In particular, putting $T = \varepsilon \log h^{-1}$ for $0 < \varepsilon \ll 1$ and $R = h^{\frac{1}{4}}$, we see that

$$R^{\frac{n-1}{2}} |\mathcal{B}|^{1/2} \leq C R^{\frac{1}{2}} h^{-\varepsilon C} (\log h^{-1} + 1) \leq C h^{\frac{1}{16}}, \qquad |\mathcal{G}| \leq C R^{1-n}.$$

Inserting this into Theorem 7.1.5 and, for simplicity, assuming that u is an eigenfunction i.e. $(-h^2 \Delta_g - 1)u = 0$, we obtain

$$\|u\|_{L^\infty(B(x,h^\delta))} \leq C_n \tau^{-\frac{1}{2}} h^{\frac{1-n}{2}} R(h)^{\frac{n-1}{2}} \left(|\mathcal{B}|^{\frac{1}{2}} \|u\|_{L^2} + \frac{|\mathcal{G}|^{\frac{1}{2}} t_0^{\frac{1}{2}}}{T^{\frac{1}{2}}} \|u\|_{L^2} \right) + C_N h^N \|u\|_{L^2}$$

$$\leq C \frac{h^{-\frac{n-1}{2}}}{\sqrt{\log h^{-1}}} \|u\|_{L^2}$$

as claimed.

8.3   Contraction

The second mechanism that we use to decompose covers of SN^*H is that of contraction. In particular, we are able to use estimates of the following form. Let $f(t) \in L^1([1, \infty))$ and suppose that such that for all $A \subset SN^*H$,

$$\mathrm{vol}_{n-1}(\varphi_t(A)) \leq f(t)\, \mathrm{vol}_{n-1}(A), \tag{8.8}$$

where vol_{n-1} denotes the Hausdorff $n-1$ measure in T^*M, to decompose good covers into *many* non-self looping sets. Note, however, that the flow-outs of these sets by φ_t *will* typically still intersect SN^*H for large times even though they do not self-intersect. One example where it is possible to verify the assumption (8.8) is when H is a subset of a stable horosphere in an Anosov manifold. In this case, the function $f(t)$ can be taken to be $Ce^{-t/C}$. We refer the reader to [CG19a, Sects. 3 and 5] for the details of the argument.

We now outline the algorithm used to construct non-self looping sets on SN^*H and hence to obtain improved estimates on the contribution to

$$\int_H u\, d\sigma_H.$$

Remark 8.3.1 One does typically expect (8.8) for all $A \subset SN^*H$. Instead, in many examples, (8.8) holds for any $A \subset B$ for some $B \subset SN^*H$. In order to obtain improved estimates on averages in this setting, it is necessary to combine the contraction mechanism illustrated in this section with the transversality mechanism illustrated in Sect. 8.2 (see [CG19a, Sect. 5]).

Choose $T_0 > 0$ such that $\|f\|_{L^1(T_0,\infty)} < \frac{1}{3}$, and define $A_1 = SN^*H$,

$$A_n := \bigcup_{t=T_0}^{2^{-n}T} \varphi_t(A_{n-1}) \cap A_{n-1}, \qquad n \geq 1$$

so that

$$\mathrm{vol}_{n-1}(A_n) \leq \tfrac{1}{3}\,\mathrm{vol}_{n-1}(A_{n-1}).$$

Therefore, if we put

$$G_j = A_{j-1} \setminus A_j, \qquad j = 1,\ldots$$

we have that G_j is $[T_0, 2^{-j}T]$ non-self looping and $\mathrm{vol}_{n-1}(G_j) \leq 3^{-j+1}\,\mathrm{vol}_{n-1}(A)$. In particular, observe that

$$\sum_j \frac{[\mathrm{vol}_{n-1}(G_j)]^{1/2}}{(2^{-j}T)^{1/2}} \leq \sum_j \frac{3^{-j+1/2}[\mathrm{vol}_{n-1}(SN^*H)]^{1/2}}{(2^{-j}T)^{1/2}} \leq C \frac{(\mathrm{vol}_{n-1}(SN^*H))^{1/2}}{T^{1/2}}.$$

In practice, we must cover A by tubes of some fixed, small radius and define the sets G_j as unions of those tubes. This can be done, and a similar construction can be used to find, for a $(\mathfrak{D}, \tau, R)$ good cover of SN^*H, $\{\mathcal{T}_j\}_{j \in \mathcal{J}}$, collections $\mathcal{G}_\ell \subset \mathcal{J}$ such that

$$|\mathcal{G}_\ell| \leq 3^{-\ell}|\mathcal{J}|, \qquad \bigcup_{j \in \mathcal{G}_\ell} \mathcal{T}_j \text{ is } [T_0, 2^{-\ell}T(R)]\text{non-self looping},$$

and hence, since $|\mathcal{J}| \leq C R^{1-n}$, so that

$$\sum_\ell \frac{|\mathcal{G}_\ell|^{1/2}}{(2^{-\ell}T(R))^{1/2}} \leq C \frac{R^{\frac{1-n}{2}}}{(T(R))^{1/2}}.$$

In practice, we must take the non-looping time $T(R) \ll \log R^{-1}$ so that, under the flow, the scale of a tube of radius R cannot change by more than some small, fixed power of R.

With the sets $\mathcal{G}_\ell$ in hand, we can then apply Theorem 7.2.2 to a eigenfunction, u of the Laplacian, with $R(h) = h^\delta$ to obtain

$$\left| \int_H u\,d\sigma_H \right| \leq C\tau^{-\frac{1}{2}}h^{\frac{1-k}{2}}R(h)^{\frac{n-1}{2}} \sum_\ell \frac{|\mathcal{G}_\ell|^{1/2}T_0^{1/2}}{2^{-\ell/2}\sqrt{\log h^{-1}}}\|u\|_{L^2} \leq C\frac{h^{\frac{1-k}{2}}}{\sqrt{\log h^{-1}}}\|u\|_{L^2},$$

which is the desired estimate.

References

[Abr70]   R. Abraham, Bumpy metrics, in *Global Analysis (Proceedings of Symposia in Pure Mathematics*, Vol. XIV. (Berkeley, California, 1968), pp. 1–3. (American Mathematical Society. Providence, R.I., 1970)

[Ano67]   D.V. Anosov, Geodesic flows on closed Riemannian manifolds of negative curvature. Trudy Mat. Inst. Steklov. **90**, 209 (1967)

[Ano82]   D.V. Anosov, Generic properties of closed geodesics. Izv. Akad. Nauk SSSR Ser. Mat. **46**(4), 675–709, 896 (1982)

[Ava56]   V.G. Avakumović, Über die Eigenfunktionen auf geschlossenen Riemannschen Mannigfaltigkeiten. Math. Z. **65**, 327–344 (1956)

[Bér77a] P.H. Bérard, On the wave equation on a compact Riemannian manifold without conjugate points. Math. Z. **155**(3), 249–276 (1977)

[Bér77b] P.H. Bérard, On the wave equation on a compact Riemannian manifold without conjugate points. Math. Z. **155**(3), 249–276 (1977)

[BHS22]   M. Blair, X. Huang, C, Sogge, Improved spectral projection estimates (2022). arXiv:2211.17266

[Bla10a] D.E. Blair, *Riemannian Geometry of Contact and Symplectic Manifolds*. Progress in Mathematics, vol. 203, 2nd edn. (Birkhäuser Boston, Inc., Boston, MA, 2010)

[Bla10b] D.E. Blair, *Riemannian Geometry of Contact and Symplectic Manifolds*. Progress in Mathematics, vol. 203, 2nd edn. (Birkhäuser Boston, Inc., Boston, MA, 2010)

[Bon17a] Y. Bonthonneau, The Θ function and the Weyl law on manifolds without conjugate points. Doc. Math. **22**, 1275–1283 (2017)

[Bon17b] Y. Bonthonneau, The Θ function and the Weyl law on manifolds without conjugate points. Doc. Math. **22**, 1275–1283 (2017)

[BS15a]   M. Blair, C. Sogge, Refined and microlocal Kakeya-Nikodym bounds for eigenfunctions in two dimensions. Anal. PDE **8**(3), 747–764 (2015)

[BS15b]   M.D. Blair, C.D. Sogge, On Kakeya-Nikodym averages, L^p-norms and lower bounds for nodal sets of eigenfunctions in higher dimensions. J. Eur. Math. Soc. (JEMS) **17**(10), 2513–2543 (2015)

[BS17a]   M.D. Blair, C.D. Sogge, Refined and microlocal Kakeya–Nikodym bounds of eigenfunctions in higher dimensions. Commun. Math. Phys. **356**(2), 501–533 (2017)

© The Editor(s) (if applicable) and The Author(s), under exclusive license to Springer Nature Switzerland AG 2023
Y. Canzani and J. Galkowski, *Geodesic Beams in Eigenfunction Analysis*, Synthesis Lectures on Mathematics & Statistics, https://doi.org/10.1007/978-3-031-31586-2

[BS17b] M.D. Blair, C.D. Sogge, Refined and microlocal Kakeya–Nikodym bounds of eigenfunctions in higher dimensions. Commun. Math. Phys. **356**(2), 501–533 (2017)

[BS18] M.D. Blair, C.D. Sogge, Concerning Toponogov's theorem and logarithmic improvement of estimates of eigenfunctions. J. Differ. Geom. **109**(2), 189–221 (2018)

[BS19] M.D. Blair, C.D. Sogge, Logarithmic improvements in L^p bounds for eigenfunctions at the critical exponent in the presence of nonpositive curvature. Invent. Math. **217**(2), 703–748 (2019)

[BW17] B. Bourgain, N. Watt, Mean square of zeta function, circle problem and divisor problem revisited (2017). arXiv:1709.04340

[CdV73] Yves Colin de Verdière, Spectre du laplacien et longueurs des géodésiques périodiques. II. Compositio Math. **27**(2), 159–184 (1973)

[CG19a] Y. Canzani, J. Galkowski, Improvements for eigenfunction averages: an application of geodesic beams (2019). arXiv:1809.06296. (to appear in J. Differential Geom.)

[CG19b] Y. Canzani, J. Galkowski, On the growth of eigenfunction averages: microlocalization and geometry. Duke Math. J. **168**(16), 2991–3055 (2019)

[CG20a] Y. Canzani, J. Galkowski, Growth of high L^p norms for eigenfunctions: an application of geodesic beams (2020). arXiv:2003.04597. (to appear in Anal. PDE)

[CG20b] Y. Canzani, J. Galkowski, Weyl remainders: an application of geodesic beams (2020). arXiv:2010.03969. (to appear in Inventiones Mathematicae)

[CG21] Y. Canzani, J. Galkowski, Eigenfunction concentration via geodesic beams. J. Reine Angew. Math. **775**, 197–257 (2021)

[CG22] Y. Canzani, J. Galkowski, Logarithmic improvements in the weyl law and exponential bounds on the number of closed geodesics are predominant (2022). arXiv:2204.11921

[CGT18] Y. Canzani, J. Galkowski, J.A. Toth, Averages of eigenfunctions over hypersurfaces. Commun. Math. Phys. **360**(2), 619–637 (2018)

[Cha74] J. Chazarain, Formule de Poisson pour les variétés riemanniennes. Invent. Math. **24**, 65–82 (1974)

[Chr16] O. Christensen, *An introduction to Frames and Riesz Bases*. Applied and Numerical Harmonic Analysis, 2nd edn. (Birkhäuser/Springer, Cham, 2016)

[CS15] X. Chen, C.D. Sogge, On integrals of eigenfunctions over geodesics. Proc. Am. Math. Soc. **143**(1), 151–161 (2015)

[DG75] J.J. Duistermaat, V.W. Guillemin, The spectrum of positive elliptic operators and periodic bicharacteristics. Invent. Math. **29**(1), 39–79 (1975)

[DG14] S. Dyatlov, C. Guillarmou, Microlocal limits of plane waves and Eisenstein functions. Ann. Sci. Éc. Norm. Supér. (4) **47**(2),371–448 (2014)

[DS99] M. Dimassi, J. Sjöstrand, *Spectral Asymptotics in the Semi-classical Limit*. London Mathematical Society Lecture Note Series, vol. 268. (Cambridge University Press, Cambridge, 1999)

[DZ16] S. Dyatlov, J. Zahl, Spectral gaps, additive energy, and a fractal uncertainty principle. Geom. Funct. Anal. **26**(4), 1011–1094 (2016)

[DZ19] S. Dyatlov, M. Zworski, *Mathematical Theory of Scattering Resonances*. Graduate Studies in Mathematics, vol. 200. (American Mathematical Society, Providence, RI, 2019)

[Gal19] J. Galkowski, Defect measures of eigenfunctions with maximal L^∞ growth. Ann. Inst. Fourier (Grenoble) **69**(4), 1757–1798 (2019)

[Goo83] A. Good, *Local Analysis of Selberg's Trace Formula*. Lecture Notes in Mathematics, vol. 1040. (Springer, Berlin, 1983)

[Gr53] Lars Gå rding, On the asymptotic distribution of the eigenvalues and eigenfunctions of elliptic differential operators. Math. Scand. **1**, 237–255 (1953)

[GT17] J. Galkowski, J.A. Toth, Eigenfunction scarring and improvements in L^∞ bounds. Anal. PDE **11**(3), 801–812 (2017)

[GWW92] C. Gordon, D. Webb, S. Wolpert, Isospectral plane domains and surfaces via Riemannian orbifolds. Invent. Math. **110**(1), 1–22 (1992)

[Hej82] D.A. Hejhal, Sur certaines séries de Dirichlet associées aux géodésiques fermées d'une surface de Riemann compacte. C. R. Acad. Sci. Paris Sér. I Math. **294**(8), 273–276 (1982)

[Hör68] L. Hörmander, The spectral function of an elliptic operator. Acta Math. **121**, 193–218 (1968)

[HT15] A. Hassell, M. Tacy, Improvement of eigenfunction estimates on manifolds of nonpositive curvature. Forum Math. **27**(3), 1435–1451 (2015)

[Hux03] M.N. Huxley. Exponential sums and lattice points. III. Proc. Lond. Math. Soc. (3), **87**(3), 591–609 (2003)

[IW21] A. Iosevich, E. Wyman, Weyl law improvement for products of spheres. Anal. Math. **47**(3), 593–612 (2021)

[Kli74] W. Klingenberg, Riemannian manifolds with geodesic flow of Anosov type. Ann. Math. **2**(99), 1–13 (1974)

[KTZ07] H. Koch, D. Tataru, M. Zworski, Semiclassical L^p estimates. Ann. Henri Poincaré **8**(5), 885–916 (2007)

[Lev52] B.M. Levitan, On the asymptotic behavior of the spectral function of a self-adjoint differential equation of the second order. Izvestiya Akad. Nauk SSSR. Ser. Mat. **16**, 325–352 (1952)

[MP49] S. Minakshisundaram, Å. Pleijel, Some properties of the eigenfunctions of the Laplace-operator on Riemannian manifolds. Canad. J. Math. **1**, 242–256 (1949)

[Ran78] B. Randol, The Riemann hypothesis for Selberg's zeta-function and the asymptotic behavior of eigenvalues of the Laplace operator. Trans. Am. Math. Soc. **236**, 209–223 (1978)

[Saf88] Y.G. Safarov, Asymptotic of the spectral function of a positive elliptic operator without the nontrap condition. Funct. Anal. Appl. **22**(3), 213–223 (1988)

[See67] R.T. Seeley, *Complex Powers of an Elliptic Operator*. (American Mathematical Society, Providence, R.I., 1967)

[Sog88] C.D. Sogge, Concerning the L^p norm of spectral clusters for second-order elliptic operators on compact manifolds. J. Funct. Anal. **77**(1), 123–138 (1988)

[STZ11] C.D. Sogge, J.A. Toth, S. Zelditch, About the blowup of quasimodes on Riemannian manifolds. J. Geom. Anal. **21**(1), 150–173 (2011)

[SV97] Y. Safarov, D. Vassiliev, *The Asymptotic Distribution of Eigenvalues of Partial Differential Operators*. Translations of Mathematical Monographs, vol. 155. (American Mathematical Society, Providence, RI, 1997). Translated from the Russian manuscript by the authors

[SXZ16] C.D. Sogge, Y. Xi, C. Zhang, Geodesic period integrals of eigenfunctions on riemann surfaces and the Gauss-Bonnet Theorem (2016). arXiv:1604.03189

[SZ99] J. Sjöstrand, M. Zworski, Asymptotic distribution of resonances for convex obstacles. Acta Math. **183**(2), 191–253 (1999)

[SZ02a] C.D. Sogge, S. Zelditch, Riemannian manifolds with maximal eigenfunction growth. Duke Math. J. **114**(3), 387–437 (2002)

[SZ02b] C.D. Sogge, S. Zelditch, Riemannian manifolds with maximal eigenfunction growth. Duke Math. J. **114**(3), 387–437 (2002)

[SZ16a] C.D. Sogge, S. Zelditch, Focal points and sup-norms of eigenfunctions. Rev. Mat. Iberoam. **32**(3), 971–994 (2016)

[SZ16b] C.D. Sogge, S. Zelditch, Focal points and sup-norms of eigenfunctions II: the two-dimensional case. Rev. Mat. Iberoam. **32**(3), 995–999 (2016)

[Vol90a] A.V. Volovoy, Improved two-term asymptotics for the eigenvalue distribution function of an elliptic operator on a compact manifold. Commun. Partial Differ. Equ. **15**(11), 1509–1563 (1990)

[Vol90b]   A.V. Volovoy, Verification of the Hamilton flow conditions associated with Weyl's con-
           jecture. Ann. Global Anal. Geom. **8**(2), 127–136 (1990)
[Wei75]   A. Weinstein, Fourier integral operators, quantization, and the spectra of Riemannian
           manifolds, in Géométrie symplectique et physique mathématique (Colloq. Internat.
           CNRS, No. 237, Aix-en-Provence, 1974), pp. 289–298. (1975). With questions by W.
           Klingenberg and K. Bleuler and replies by the author
[Wey12]   H. Weyl, Das asymptotische Verteilungsgesetz der Eigenwerte linearer partieller Differ-
           entialgleichungen (mit einer Anwendung auf die Theorie der Hohlraumstrahlung). Math.
           Ann. **71**(4), 441–479 (1912)
[Wym17]   E.L. Wyman, Integrals of eigenfunctions over curves in compact 2-dimensional manifolds
           of nonpositive sectional curvature (2017). arXiv:1702.03552
[Wym18]   E.L. Wyman, Period integrals in non-positively curved manifolds (2018).
           arXiv:1806.01424
[Wym19]   E.L. Wyman, Looping directions and integrals of eigenfunctions over submanifolds. J.
           Geom. Anal. **29**(2), 1302–1319 (2019)
[Wym20]   E.L. Wyman, Explicit bounds on integrals of eigenfunctions over curves in surfaces of
           nonpositive curvature. J. Geom. Anal. **30**(3), 3204–3232 (2020)
[Zel92]   S. Zelditch, Kuznecov sum formulae and Szegő limit formulae on manifolds. Commun.
           Partial Differ. Equ. **17**(1–2), 221–260 (1992)
[Zwo12]   M. Zworski, *Semiclassical Analysis*. Graduate Studies in Mathematics, vol. 138. (Amer-
           ican Mathematical Society, Providence, RI, 2012)

MIX
Papier aus verantwortungsvollen Quellen
Paper from responsible sources
FSC® C105338

If you have any concerns about our products,
you can contact us on
ProductSafety@springernature.com

In case Publisher is established outside the EU,
the EU authorized representative is:
**Springer Nature Customer Service Center GmbH
Europaplatz 3, 69115 Heidelberg, Germany**

Printed by Libri Plureos GmbH
in Hamburg, Germany